Semiconductor and Integrated Circuit Fabrication Techniques

PETER E. GISE
RICHARD BLANCHARD

Prepared by
The Fairchild Management and
Career Development Center

Semiconductor and Integrated Circuit Fabrication Techniques

RESTON PUBLISHING COMPANY, INC.
A Prentice-Hall Company
Reston, Virginia

Library of Congress Cataloging in Publication Data
Fairchild Camera and Instrument Corporation.
  Semiconductor processing.

  Includes bibliographical references and index.
  1. Semiconductors. I. Title.
TK7871.85.F298   1979   621.3815′2   78-18373
ISBN 0-87909-668-3

© 1979 by Reston Publishing Company, Inc.
A Prentice-Hall Company
Reston, Virginia

1   3   5   7   9   10   8   6   4   2

PRINTED IN THE UNITED STATES OF AMERICA

Contents

Preface

The purposes of this text are to provide a single source of reference to those individuals involved in the processing of semiconductors, and to introduce students of other technologies to the technology of semiconductor processing. The text arose from a sequence of college courses for the semiconductor technician, and was expanded to include many aspects of process design of interest to the processing engineer.

The more complex lessons in the text are approached at two levels of detail. The first level covers the basics of the particular topic and terminates at the technician level. The second level covers more advanced material and should be used at the engineering level.

Each lesson should take an average of two hours to complete. For such self-paced study audio cassettes are available. The set may be purchased from Fairchild Camera & Instrument Corporation, Corporate Training, Mountain View, California.

Semiconductor and Integrated Circuit Fabrication Techniques

1

Semiconductor Physics I

1-1 ATOMIC STRUCTURE

Early structural models of the atom pictured it as having a nucleus composed of positively charged protons and electrically neutral neutrons, surrounded by orbitals or shells containing negatively charged electrons. (Figure 1-1). This model of the atom is being continuously refined by atomic physicists, but the features of the model are sufficient to explain many of the physical phenomena observed in many materials, including most semiconductors.

An atom that has the same number of electrons and protons is electrically neutral. However, the gain or loss of electrons from the orbitals surrounding the nucleus produces an atom that is charged either positively or negatively. An atom charged in such a fashion is called an ionized atom or an ion. The majority of the physical and chemical properties of an atom are determined by the number of electrons in the outermost orbital, since these electrons are the means by which the atom interacts with the outside world.

All atoms with the same number of protons (regardless of the number of neutrons or electrons) are the same element. Unionized atoms with the same number of protons must also have the same number of electrons. Hence, only the number of neutrons contained in the nucleus can differ. Atoms with the same number of protons but a different number of neutrons are isotopes of the element.

Studies by several nineteenth century chemists detected similarities in the physical and chemical properties of elements having

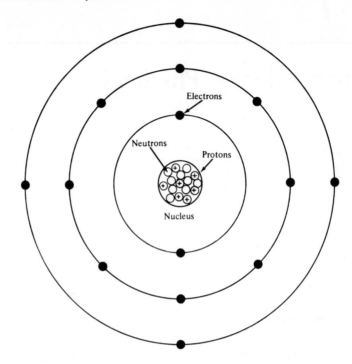

Figure 1-1 A silicon atom.

different densities. Grouping of elements with similar properties based on their densities led to the Periodic Table (Figure 1-2). This table was arrived at largely by experimental means, but it offers large insights into the behavior of materials. The uncomplicated but accurate picture of semiconductors obtained through use of the Periodic Table is sufficient for all but advanced work in semiconductors.

The Periodic Table (modified from Mendeleev's) is based upon the arrangement of electrons around the nucleus of an atom. Every atom has orbitals or shells that can be occupied by electrons. The orbitals closer to the nucleus can hold fewer electrons than the orbitals farther away. Electrons fill orbitals starting from the innermost. The rows of elements in the Periodic Table correspond to the filling of an orbital with electrons. When an orbital is filled, a new row in the Periodic Table is begun. Elements that are in the same column in the Periodic Table have the same number of electrons in the outermost orbital. The columns are given "group numbers" which tell the number of electrons in the outer orbital. (We will now concentrate on the Group I elements.)

Atomic weights are the most recent adopted by the International Union of Chemistry: none are given for artificially produced elements.

GROUP		I	II	III	IV	V	VI	VII	VIII			O
Period	Series											
1	1	1H 1.0080										2He 4.003
2	2	3Li 6.940	4Be 9.013	5B 10.82	6C 12.010	7N 14.008	8O 16.000	9F 19.00				10Ne 20.183
3	3	11Na 22.997	12Mg 24.32	13Al 26.97	14Si 28.06	15P 30.98	16S 32.066	17Cl 35.457				18A 39.944
4	4	19K 39.096	20Ca 40.08	21Sc 45.10	22Ti 47.90	23V 50.95	24Cr 52.01	25Mn 54.93	26Fe 55.85	27Co 58.94	28Ni 58.69	
	5	29Cu 63.54	30Zn 65.38	31Ga 69.72	32Ge 72.60	33As 74.91	34Se 78.96	35Br 79.916				36Kr 83.7
5	6	37Rb 85.48	38Sr 87.63	39Y 88.92	40Zr 91.22	41Nb 92.91	42Mo 95.95	43Tc	44Ru 101.7	45Rh 102.91	46Pd 106.7	
	7	47Ag 107.880	48Cd 112.41	49In 114.76	50Sn 118.70	51Sb 121.76	52Te 127.61	53I 126.92				54Xe 131.3
6	8	55Cs 132.91	56Ba 137.36	6 57-71 Rare earths*	72Hf 178.6	73Ta 180.88	74W 183.92	75Re 186.31	76Os 190.2	77Ir 193.1	78Pt 195.23	
	9	79Au 197.2	80Hg 200.61	81Tl 204.39	82Pb 207.21	83Bi 209.00	84Po 210	85At				86Rn 222
7	10	87Fr	88Ra 226.05	89 Actinide series**								

*Rare earths: 57La 132.92 | 58Ce 140.13 | 59Pr 140.92 | 60Nd 144.27 | 61Pm | 62Sm 150.43 | 63Eu 152.0 | 64Gd 156.9 | 65Tb 159.2 | 66Dy 162.46 | 67Ho 164.94 | 68Er 167.2 | 69Tm 169.4 | 70Yb 173.04 | 71Lu 174.99

**Actinide series: 89Ac 227 | 90Th 232.12 | 91Pa 231 | 92U 238.07 | 93Np | 94Pu | 95Am | 96Cm | 97Bk | 98Cf

Figure 1-2 Periodic table of the elements.

Dimitri Ivanovich Mendeleev, the Russian chemist who devised the Periodic Table, noted that atoms with eight electrons in their outer orbitals are chemically inert. The observation that atoms have a complete set of electrons when they have eight electrons in the outer shell, explains the way elements reacted to form compounds. Group I elements (with one electron in their outer shell) react with Group VII elements (with seven electrons in their outer shell). The Group VII atom "borrows" the electron to complete its outer shell, leaving the Group I atom with zero electrons in its outer shell, but with completed shells beneath. Each element then has a full complement of electrons in its outer shell. The atoms are held together by the electric force between the atoms with one extra electron and the atom with one less electron. This type of bonding is known as ionic bonding.

If a Group II and a Group VI element combine, each atom satisfies its need for electrons. However, the Group VI atom has difficulty capturing the extra electrons, so it shares them instead. The bond formed is less ionic (electron-taking) and more covalent (electron-sharing). In a similar fashion, Group III atoms combine with Group V atoms, and group IV atoms combine with Group IV atoms. A Group IV atom will share one of its four electrons with each of its four nearest neighbors part of the time, and borrow one electron from its neighbors part of the time.

1-2   CLASSIFICATION OF MATERIALS

One method used by scientists to classify materials is to group them by their ability to conduct electricity. Three broad classifications of materials are:

1. Insulator—does not conduct electricity to an appreciable degree.
2. Metal—conducts electricity easily.
3. Semiconductor—conducts electricity poorly when pure.

If we look at the electron structure of these three classifications of materials (Figure 1-3), we see that:

1. Insulators have all electrons tightly bound, so none are free to carry current.
2. Metals have many electrons readily available to carry current.
3. Semiconductors have some electrons free to carry current.

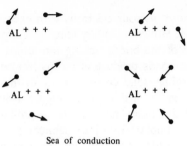

Sea of conduction
electrons free to move

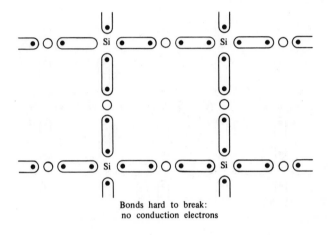

Bonds hard to break:
no conduction electrons

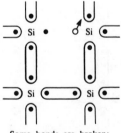

Some bonds are broken:
few conduction electrons
and holes result

Figure 1-3   Bonding diagrams of a metal, an insulator, and a semiconductor.

Taking a closer look at semiconductors, of which silicon and germanium are the most widely used, we see that they both belong to Group IV in the Periodic Table. When these elements are crystalline, an atom shares one of its four electrons with each of its nearest neighbors (Figure 1-4a). However, at any temperature greater than absolute zero (0°K), some of the bonds linking the atoms are broken (Figure 1-4b). The broken bonds produce electrons free to conduct electricity. In addition, the broken bond corresponding to the absence of an electron is also free to move in the lattice (Figure 1-4c). (The absence of an electron is called a hole; this concept is similar to that calling the absence of water a bubble.) In a pure semiconductor crystal, the number of broken bonds depends only on the temperature. Since every broken bond produces both a hole and an electron, they are present in

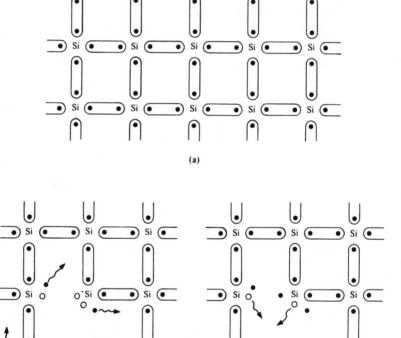

Figure 1-4 **(a)** Silicon at absolute zero; **(b)** electron conduction in silicon; **(c)** hole conduction in silicon.

equal numbers. The symbol n is used to signify the number of electrons/cm³ in a semiconductor, while the symbol p signifies the number of holes/cm³ in a semiconductor. Since they are equal in pure or "intrinsic" silicon, we can say that n equals p. The number of broken bonds in an intrinsic sample is called n_i, and it follows that

$$n = p = n_i, \quad \text{and} \tag{1-1}$$

$$n \cdot p = n_i^2 \tag{1-2}$$

where n_i^2 depends only on temperature. In silicon at room temperature (27° C) $n_i = 1.4 \times 10^{10}/\text{cm}^3$, and $n_i^2 \cong 2 \times 10^{20}/\text{cm}^6$.

The presence of equal numbers of holes and electrons leads to no interesting phenomena, but the ability to increase the number of holes or electrons by adding trace amounts of impurities called dopants, means that regions of semiconductor materials can be altered to perform useful functions. Silicon has four electrons in the outer shell which it shares with its four nearest neighbors. The substitution for silicon of an atom from Group V, for example phosphorus, results in the phosphorus sharing one of its five electrons with each of its four nearest neighbors (Figure 1-5a). The extra electron is not needed for bonding purposes, and is free to conduct electrical current. Semiconductors containing an excess of conduction electrons are called n-type. In an analogous manner, additional holes can be provided by substituting an atom like boron for a silicon atom (Figure 1-5b). Semiconductors containing an excess of holes are called p-type. Atoms supplying additional electrons for the conduction process are called donors; the number of donors/cm³ in a semiconductor is N_D. Atoms that supply additional holes for the coduction process are called acceptors; the number of acceptors/cm³ in a semiconductor is N_A. For silicon, potential donor atoms are the atoms of Group V with five electrons in their outer shell.

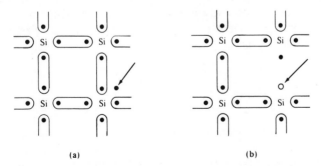

(a)   (b)

Figure 1-5   (a) Extra electron easy to remove; (b) extra hole easy to remove.

The atoms frequently used to dope silicon n-type are phosphorus, arsenic, and antimony. Potential acceptors for silicon are the Group III atoms with three electrons in their outer shell. The atoms used to dope silicon p-type are boron, aluminum, and gallium (boron is used most frequently).

An increase in the number of conduction electrons present in a semiconductor causes a corresponding decrease in the number of holes, and vice versa. The equation $n \cdot p = n_i^2$ is valid even when n does not equal p ($n \neq p$). If only donor atoms have been added to silicon and the number of donors is less than $10^{19}/cm^3$, ($N_D < 10^{19}/cm^3$), all of the donors produce conduction electrons. It follows that, in this case,

$n = N_D$, and $p = \dfrac{n_i^2}{N_D}$. In a similar fashion, if only acceptors are added to a bar of silicon, and their number is less than $10^{19}/cm^3$, ($N_A < 10^{19}/cm^3$), each acceptor atom produces one hole. In this case, $p = N_A$, and

$n = \dfrac{n_i^2}{N_A}$.

When both donors and acceptors are added to a semiconductor, they tend to cancel each other out. When more donor than acceptor atoms are added, ($N_A < N_D$), the donor atoms cancel out the effect of all of the acceptor atoms, and the number of electrons is the difference between the number of donors and the number of acceptors ($n = N_D - N_A$). In an analogous manner, if more acceptor atoms are added than donor atoms, the acceptor atoms cancel out the effect of all of the donor atoms, and the number of holes is the difference between the number of acceptors and the number of donors ($p = N_A - N_D$). In both cases, the product of $n \cdot p$ remains constant, so the carrier type in the minority can be determined using the formula $n \cdot p = n_i^2$.

The amount of dopant present in a semiconductor is determined by measuring its conductivity or resistivity. The resistivity of a material is the opposing force a material has to a voltage placed across it. The symbol for resistivity is the Greek letter ρ. The units of resistivity are ohm-centimeters (Ω-cm). The conductivity is related to the resistivity by the equation

$$\sigma = \frac{1}{\rho} \qquad (1\text{-}3)$$

The conductivity of a sample depends upon the number of free carriers (holes and/or electrons) and their mobility or the ease with which they move through the sample. If the resistivity (or the conductivity) of a material is known, the resistance of a box-shaped piece of material is determined by the formula:

$$R = \frac{\rho L}{A} \qquad (1\text{-}4)$$

where R = the resistance of the material (units of ohms)
        L = the length of the material from contact to contact
        A = the cross-sectional area of the material (area = height × width)

The resistance of a piece of material is related to the applied voltage (V) and the current that flows (I) by the equation:

$$V = RI \text{ or } R = \frac{V}{I} \qquad (1\text{-}5)$$

In a semiconductor (and in other industrial materials as well) the "sheet resistance" of a material is an often-measured parameter. The symbol for sheet resistance is R_s. Sheet resistance is measured in ohms per square ($\Omega/\square$). The resistance of a resistor made up of n squares laid in a row is nR_s. (For instance, if 10 squares of material are laid in a row with $R_s = 100 \ \Omega/\square$, $nR_s = 10R_s = 1000 \ \Omega$.) Sheet resistance is measured using a four-point probe. (Figure 1-6.). The formula relating sheet resistance to current and voltage is:

$$R_s = 4.53 \ \frac{V}{I} \qquad (1\text{-}6)$$

This equation is valid when:

1. The thickness of the layer being measured is much less than the spacing between the probes, and
2. The size of the piece of material being measured is much greater in length and width than the probe spacing.

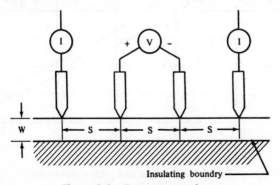

Figure 1-6   Four-point probe.

If a thin layer of material is uniformly doped and the sheet resistance is known, the resistivity ρ is found with the equation:

$$\rho = R_s \times \text{thickness} \qquad \text{or} \qquad \rho = R_s \cdot t \qquad (1\text{-}7)$$

If the thickness of the sample is much greater than the probe spacing, the formula

$$\rho = 2\pi S \frac{V}{I} \qquad (\pi = 3.14159) \qquad (1\text{-}8)$$

relates the current and voltage readings of a four-point probe to the resistivity of the material, where S is the spacing between the probes on the four-point probe.

The resistivity of silicon depends upon the number of acceptor and donor atoms added and the temperature. When only acceptor or donor atoms have been added to a bar of silicon, the resistivity of the silicon can be obtained from Figure 1-7. Conversely, the doping concentration of a uniformly doped sample can be determined if the resistivity is known. To determine the resistivity of a sample if the doping concentration and type is known, find the doping concentration along the bottom of the graph, then proceed upward until the line corresponding to the type of dopant (p-type or n-type) is encountered. Then proceed horizontally to the left to obtain the resistivity. The inverse operation is performed to obtain the doping concentration from the resistivity.

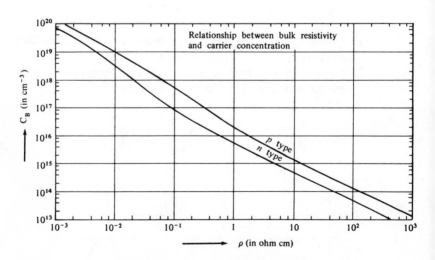

Figure 1-7 n-type and p-type resistivity.

REVIEW EXERCISES: SEMICONDUCTOR PHYSICS I

1. A sample of silicon is doped with 10^{15} atoms/cm³ of phosphorus
   a. Determine the donor concentration N_D,
   b. Determine the acceptor concentration N_A,
   c. Determine the electron concentration n,
   d. Determine the hole concentration p
   e. Determine the resistivity ρ

2. A sample of silicon is doped with 2×10^{16} atoms/cm³ of boron.
   a. Determine the donor concentration N_D,
   b. Determine the acceptor concentration N_A,
   c. Determine the electron concentration n,
   d. Determine the hole concentration p,
   e. Determine the resistivity ρ

3. If a sample of silicon is doped with 3×10^{17} atoms of arsenic and 5×10^{17} atoms of boron,
   a. Determine the donor concentration N_D,
   b. Determine the acceptor concentration N_A,
   c. Determine the electron concentration n,
   d. Determine the hole concentration p.

4. A 4-point probe measurement has been made on a sample resulting in:

$$V = 5 \times 10^{-3} \text{ volts}$$
$$I = 4.5 \times 10^{-3} \text{ amps}$$

What is the sheet resistance R_s of the sample?

5. A sample of material has the following properties:
   its length is 100 microns;
   its width is 5 microns;
   its height is 2 microns;
   its resistivity is 2 ohm-centimeters.

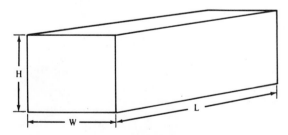

Determine the resistance of the bar of material.

6. A sample of germanium is uniformly doped with boron atoms to produce an impurity concentration of 5×10^{16} atoms/cm^3. If the intrinsic carrier concentration at 300°K is 2.43×10^{13} carriers/cm^3, determine the hole and electron concentrations for this sample.

7. If the temperature of the sample in problem 6 is increased, the imbalance between the majority and minority carrier concentrations decreases. If the intrinsic carrier concentration increases exponentially with temperature at the rate of 6% per °K, at what temperature is the minority carrier concentration equal to 2% of the majority carrier concentration?

8. An unknown semiconductor material has a hole concentration of 10^{15} carriers/cm^3 and an electron concentration of 4×10^{13} carriers/cm^3. Determine the intrinsic carrier concentration and the net impurity concentration.

9. A silicon sample is doped with 2×10^{16} acceptors/cm^3 and 5×10^{15} donors/cm^3. What type of impurity and in what concentration should be added to make the equilibrium electron and hole concentrations the same at room temperature?

10. A silicon bar with a length of 1 cm and a height and width of 0.1 cm has a resistance measured from end to end of 10 ohms. If a hot probe measurement indicates the bar to be n-type, determine the donor concentration.

2

Semiconductor Physics II

The classification of materials into metals, insulators, and semiconductors can be viewed in a way different from that used in the first part of this presentation. The allowable orbitals or shells that electrons can occupy correspond to allowable energy levels for the electrons. As atoms are brought together, these energy levels enlarge to form allowable energy bands for electrons. An insulator can be viewed as having a completely full energy band separated by a large energy gap from the next allowable band (Figure 2-1a). Similarly, a metal can be viewed as having two overlapping energy bands, allowing electrons to move easily to a higher energy level to carry current (Figure 2-1b). A semiconductor can be seen as having two energy levels separated by only a narrow energy band (Figure 2-1c). The energy present in the crystal is sufficient to cause a few of the electrons in the lower (valence) energy band to jump to the upper (conduction) energy band (Figure 2-2). This jump results in electrons in the conduction band and holes in the valence band.

The addition of donor or acceptor dopants to a semiconductor adds impurity energy levels in the energy band gap near the conduction and valence band. Donor atoms produce electrons, so the donor impurity level is just below the conduction band (Figure 2-3). Acceptor atoms introduce levels that can be occupied by electrons just above the valence band. The occupation of these levels by electrons from the valence band results in holes (Figure 2-4). When both donor and acceptor atoms are added to a semiconductor, electrons from the donor level occupy the acceptor level until it is filled, or until all of the donor atoms have left the donor level.

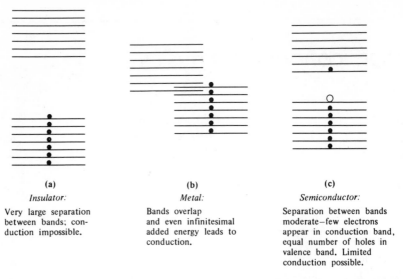

(a)	(b)	(c)
Insulator:	*Metal:*	*Semiconductor:*
Very large separation between bands; conduction impossible.	Bands overlap and even infinitesimal added energy leads to conduction.	Separation between bands moderate—few electrons appear in conduction band, equal number of holes in valence band. Limited conduction possible.

Figure 2-1   Energy-level diagram of **(a)** insulator **(b)** metal and **(c)** semiconductor.

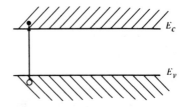

Figure 2-2   Transition of an electron from valence band to conduction band.

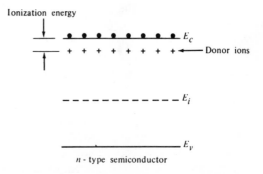

Figure 2-3   Donor levels in semiconductors.

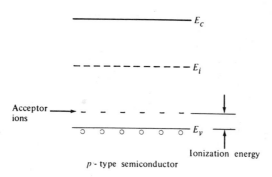

Figure 2-4   Acceptor levels in semiconductors.

The band model of semiconductors complements the bond model discussed previously. Familiarity with both of these models aids in understanding the various phenomena encountered in the semiconductor field.

2-1   RESISTIVITY

The addition of impurities to a semiconductor modifies the resistivity in a manner that produces useful behavior in semiconductors. The conductivity of a material depends on the number of holes and electrons, the charge each particle carries (called q, the charge of an electron = 1.6×10^{-19} coulombs), and the ease with which the holes and electrons move through the material. The formula for the conductivity of a material can be written as:

$$\sigma = qn\mu_n + qp\mu_p \tag{2-1}$$

where μ_p = hole mobility
              μ_n = electron mobility

The terms n and p are determined as previously discussed. The ease with which carriers (either holes or electrons) move through a crystal is influenced by the total number of impurity atoms present. Each impurity atom in the lattice is a slight disruption in the otherwise regular crystal structure. A graph of the mobility of both holes and electrons in silicon at 27°C is shown in Figure 2-5. The total dopant concentration C_T is the sum of the number of donor and acceptor atoms.

$$C_T = N_A + N_D \tag{2-2}$$

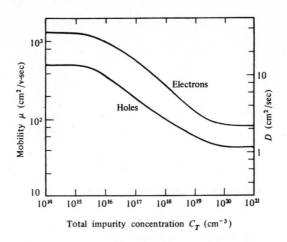

Figure 2-5   Electron and hole mobilities in silicon.

The graph of resistivity versus impurity concentration shown in Figure 1-7 is valid when only acceptor or donor atoms are added to silicon. If both types of impurities are added, the resistivity of the material must be calculated.

Example 1 $N_D = 2 \times 10^{15}/\text{cm}^3$; $N_A = 4 \times 10^{15}/\text{cm}^3$ at 27°C; determine the resistivity of the sample.
First, determine n and p.

$$p = N_A - N_D = 2 \times 10^{15}/\text{cm}^3$$

$$n = \frac{n_i^2}{p} = 1 \times 10^5$$

μ_n and μ_p are found by noting that $N_D + N_A$
$= 6 \times 10^{15}/\text{cm}^3$

$\mu_n = 1100$ cm²/v-sec; $\mu_p = 400$ cm²/v-sec

$$\sigma = q\,(\mu_n n + \mu_p p) \cong q\mu_p p$$

now

$$\rho = \frac{1}{\sigma} = \frac{1}{(1.6 \times 10^{-19})\,(400)\,(2 \times 10^{15})} = 7.8\ \Omega\text{-cm}$$

Example 2 $N_D = 6 \times 10^{17}/\text{cm}^3$; $N_A = 3 \times 10^{17}/\text{cm}^3$ at 27°C; determine the resistivity of the sample.

First, determine n and p.

$$n = N_D - N_A = 3 \times 10^{17}/\text{cm}^3$$

$$p = \frac{n_i^2}{n} = 6.7 \times 10^2/\text{cm}^3$$

μ_n and μ_p are found by first determining $C_T = N_A + N_D$; thus $\mu_n \cong 700$, $\mu_p = 200$.

$$\sigma = q\,(\mu_n n + \mu_p p) \cong q\mu_n n = (1.6 \times 10^{-19}) \\ (700)\,(3 \times 10^{17})$$

$$\sigma = 3.36 \times 10^1 = 33.6\,\frac{1}{\Omega\text{-cm}}$$

$$\rho = .028\ \Omega\text{-cm}$$

2-2 CARRIER TRANSPORT

The carriers present in a semiconductor move by one of two processes, drift or diffusion. Drift is the motion caused by the presence of an electric field. With no applied field, carriers move about randomly in a semiconductor (Figure 2-6a). Under the influence of an applied field, the carriers acquire a directed component of motion (Figure 2-6b). The sum of the directed components of drift result in current flow in the sample. Carriers also move by the process of diffusion. Diffusion is the motion of particles from regions of high concentration to regions of low concentration caused by random motion. The result of both types of carrier motion determines the total current flow in a material.

The diffusion of the mobile carrier (either holes in p-type silicon or electrons in n-type silicon) from higher temperature regions to lower temperature regions can be used to determine whether a sample of silicon is n-type or p-type. If an area of a silicon wafer is heated locally, as shown in Figure 2-7, the majority carriers diffuse away from the hot region, a voltage is then produced, and can be measured to determine the conductivity of the sample. If the sample is n-type, the voltage on the hot probe is positive with respect to a second probe. Similarly, if the sample is p-type, the voltage on the hot probe is negative with respect to a second probe. This testing technique is useful on samples unless there is just a thin layer of oppositely doped material on the surface of the wafer.

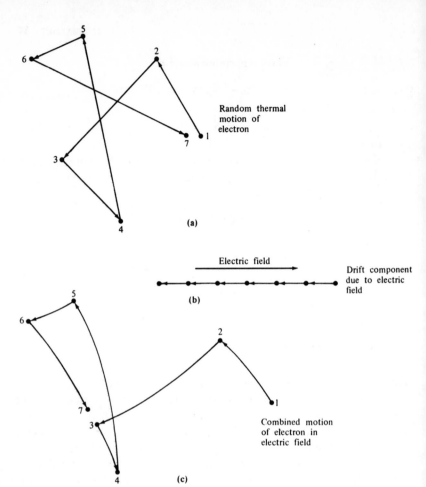

(a)

Random thermal
motion of
electron

Electric field

Drift component
due to electric
field

(b)

Combined motion
of electron in
electric field

(c)

Figure 2-6 Drift under the influence of an electric field.

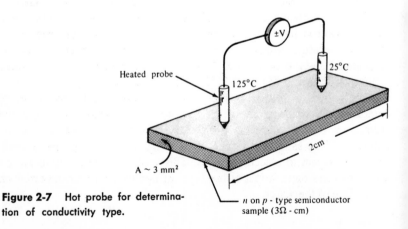

±V

Heated probe

125°C

25°C

2cm

A ~ 3 mm²

Figure 2-7 Hot probe for determination of conductivity type.

n on p - type semiconductor
sample (3Ω - cm)

REVIEW EXERCISES: SEMICONDUCTOR PHYSICS II

1. A bar of silicon is doped with 2×10^{15} arsenic atoms. Calculate the resistivity of this bar, and compare your answer with Figure 1-7.

2. A bar of silicon contains 1×10^{18} boron atoms/cm^3, and 3×10^{18} antimony atoms/cm^3.
   a. Determine the donor and acceptor concentrations (N_D and N_A).
   b. Determine the hole and electron concentrations (p and n).
   c. Determine the mobilities of the holes and the electrons (μ_p and μ_n).
   d. Determine the resistivity of the bar.
   e. Why does the answer from (**d**) differ from that obtained from Figure 1-7, with $N_D = 3 \times 10^{18}$/cm^3?

3. An n-type silicon cube 1.0 cm on a side is doped with 1×10^{14} donors/cm^3. A smaller p-type region, measuring 0.5 cm on a side, is diffused into the center of the top surface of the cube. If the resistivity of the p-region is 2.5 Ω −cm, find the acceptor concentration in the p-region and the total number of impurity atoms in the p-region.

4. Draw the energy level diagram for silicon in equilibrium at 27°C and doped with 3×10^{17} phosphorus atoms/cm^3 and 2.9×10^{17} boron atoms/cm^3. Are all the impurities ionized?

5. When two opposite type impurities are added to a semiconductor such that the number of donors equals the number of acceptors the crystal is said to be compensated. Is the resulting crystal electrically intrinsic? Explain.

6. Under what conditions are the following expressions valid?
   a. $np = n_i^2$
   b. $p + N_D = n + N_A$.

7. A piece of silicon is doped with 7×10^{15} boron atoms/cm^3 and 3×10^{15} phosphorous atoms/cm^3. Find the electron and hole concentrations at 27°C.

8. Given a silicon sample 1 cm on a side at 27°C, find the resistance between any two faces.

3

Wafer Preparation I

The silicon used in the fabrication of semiconductor devices is very pure and is in the shape of flat, circular wafers. Each wafer of silicon is a single crystal. The preparation of these wafers prior to the actual fabrication sequence that produces electrical devices is a complex procedure that is a story in itself.

Silicon is a very abundant material on the Earth's surface. However, it is found in the form of compounds, and must be separated from other elements before it can be used. Sand available in many locations is silicon dioxide (SiO_2) containing less than 1% impurities. This sand is used as the starting point for the manufacture of silicon wafers. The following steps are followed in the production of ultrapure silicon for semiconductor devices.

Step 1. Silicon dioxide (sand) is mixed with carbon and reacted to form silicon (99% pure) and carbon dioxide.

$$SiO_2 + C \rightarrow Si + CO_2\uparrow \tag{3-1}$$

This reaction produces both silicon and a gaseous byproduct that is easily exhausted. The silicon following this reaction is 99% pure; far from the quality of silicon required by contemporary technology. Further steps are necessary to remove additional unwanted impurities.

Step 2. Silicon is reacted with hydrogen chloride to form tri-chlorosilane.

$$Si + 3HCl \rightarrow SiHCl_3 + H_2\uparrow \qquad (3\text{-}2)$$

The production of trichlorosilane leaves behind the unwanted impurities, in many cases yielding a chemical with sufficient purity for semiconductor devices. If higher purity is desired, the trichlorosilane may be distilled.

Step 3. The trichlorosilane is decomposed using electric current in a chamber with a controlled atmosphere, producing rods of ultrapure polycrystalline silicon.

$$SiHCl_3 + H_2 \rightarrow Si + 3HCl \qquad (3\text{-}3)$$

The polycrystalline silicon (silicon containing many crystals) is now ready for the crystal-growing process.

3-1 SILICON CRYSTAL GROWTH

Two methods are presently used to grow single-crystal silicon. These two methods are called *Czochralski* and *float zone* crystal growth, respectively. (These two crystal growth methods are often abbreviated as CZ and FZ). Czochralski crystal growth utilizes a crucible in which pieces of polycrystalline silicon have been heated to their melting point of 1415°C (Figure 3-1). The crucible containing the silicon is made of quartz (SiO_2) and is heated by either induction (RF) or thermal resistance methods. The crucible rotates during the growth process to prevent the formation of local hot or cold regions. The atmosphere around the crystal-growing apparatus or crystal puller is controlled to prevent contamination of the molten silicon. Argon is often used as the ambient gas. When the temperature of the silicon has stabilized, an arm with a piece of silicon mounted on the end is slowly lowered until it comes into contact with the surface of the molten silicon. This piece of silicon is called the seed crystal, and is the starting point for the subsequent growth of a much larger crystal. As the bottom of the seed crystal begins to melt in the molten silicon, the downward motion of the rod holding the silicon is reversed. As the seed crystal is slowly withdrawn from the melt, the molten silicon adhering to the crystal freezes or solidifies, taking on the crystal structure of the seed crystal. The rod continues its upward movement, forming an even larger crystal. The

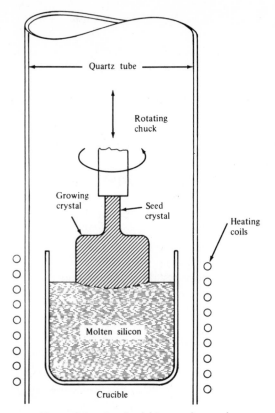

Figure 3-1 Czochralski crystal growth.

crystal growth terminates when the silicon in the crucible is depleted. By carefully controlling the temperature of the crucible and the rotation speeds of the crucible and the rod, precise control of the diameter of the crystal is maintained. The desired impurity concentration is obtained by adding the impurities to the melt in the form of heavily doped silicon prior to crystal growth.

Float zone crystal growth proceeds directly from the rod of polycrystalline silicon previously obtained. A rod of silicon of the appropriate diameter is held at the top and placed in the crystal-growth chamber. A single seed crystal is clamped beneath, in contact with the other end of the polycrystalline rod (Figure 3-2). The rod is enclosed in a chamber with a controlled atmosphere and an induction heating coil is placed around it. The coil melts a small length of the rod starting with part of the single seed crystal. The molten zone is then slowly moved upward along the length of the rod by moving the coil upward.

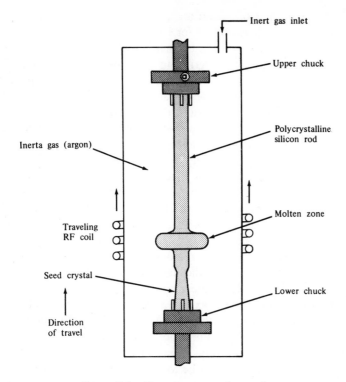

Figure 3-2   Float zone crystal growth.

The molten region that freezes first is in contact with the single seed crystal. This region assumes the crystal structure of the seed. As the molten zone proceeds along the length of the rod, the polycrystalline rod melts and then freezes, becoming a single crystal rod of silicon. The diameter of the crystal is controlled by the motion of the heating coil. The desired impurity level is obtained by starting with polycrystalline silicon doped to the appropriate level.

The single crystal grown by either the Czochralski or the float zone technique is now ready to be sliced into wafers.

3-2   WAFER ORIENTATION, SAWING, AND POLISHING

The silicon crystals are first ground perfectly round (if necessary), and the rotational orientation of the crystal is ascertained. The seed crystal has determined which crystal face will be present on the wafer surface, but the rotational position of the rod determines other axes of the crystal. Since the bar of silicon is one crystal, it has preferential break or cleavage planes. It is critical for later device separation to align the circuits precisely with respect to their cleavage planes. This precise alignment is accomplished by grinding a flat along the crystal (prior to sawing) that is used as a reference during all subsequent processing steps. X-ray diffraction provides a fast and accurate method of determining the crystal orientation prior to grinding the flat.

The silicon crystal is then sawed into thin slices called wafers. Extreme care is taken to minimize the amount of the single crystal silicon that is lost in the sawing process by using the inside diameter of a ring-shaped saw blade. The blade is coated with diamond powder to enable it to cut through the hard silicon. The sawing process leaves wafers with saw marks on both sides that must be removed. A silicon etchant is used to remove the saw marks and any accompanying damage from both sides of the wafer.

Care must be taken to remove any crystal damage introduced by the sawing operations, or the damage may prevent the successful fabrication of devices. The wafers are next mounted on large circular polishing plates using either wax or a vacuum to hold them. The polishing plates are mounted on a polisher, and one side of the wafer receives a mirrorlike finish. The polishing operation uses a polishing solution that simultaneously chemically etches and mechanically polishes the wafers. The polishing pad must be tough and durable. When the wafers have reached the proper thickness range and surface quality, the polishing plates are removed and the wafers are dismounted. The wafers are thoroughly cleaned to remove any residual contamination, and inspected to insure that wafers with imperfect surfaces are not shipped. Wafers that pass the final inspection are ready to start on their journey to become devices.

REVIEW EXERCISES: WAFER PREPARATION I

1. Why is it desirable to end up with a solid and a gas—or a liquid and a gas—following a chemical reaction?

2. **a.** What contaminant may CZ silicon contain that FZ silicon may not contain?

   b. Where does this contaminant come from?

3. Does silicon or silicon dioxide have a higher melting point? Why?

4. a. Why is the crystal orientation of a wafer important?
   b. How is the orientation denoted?
5. Define the term polysilicon.
6. What is the purpose of the argon gas during crystal growth?
7. Why is a seed crystal used for crystal growth?
8. What two variables are used to control the diameter of the silicon rod?

4

Wafer Preparation II

4-1 CRYSTAL ORIENTATION

The orientation of a silicon crystal is an important parameter in the device fabrication sequence. One method used to describe the orientation of crystals is through the use of Miller indices. The Miller indices of a plane of silicon are determined by the point at which the plane intersects the x, y, and z axes, as shown in Figure 4-1. To determine the Miller indices of a plane from the points of intersection, the following procedure is used:

$$x \text{ index} = \frac{1}{x \text{ point of intersection}} \tag{4-1}$$

$$y \text{ index} = \frac{1}{y \text{ point of intersection}} \tag{4-2}$$

$$z \text{ index} = \frac{1}{z \text{ point of intersection}} \tag{4-3}$$

If we consider the plane intersecting the set of axes in Figure 4-1, we can determine the Miller indices of the plane. This plane intersects the axes at $x = 1$, $y = 1$, and $z = 1$; the Miller indices are:

$$x \text{ index} = \frac{1}{x \text{ point of intersection}} = \frac{1}{1} = 1$$

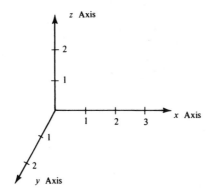

Figure 4-1   The crystal axis set.

$$y \text{ index} = \frac{1}{y \text{ point of intersection}} = \frac{1}{1} = 1$$

$$z \text{ index} = \frac{1}{z \text{ point of intersection}} = \frac{1}{1} = 1$$

This plane is a <111> plane of the crystal. Other planes parallel to this one are also <111> planes. The <111> planes intersect the crystal axes so that a triangle is formed by their intersection. Crystals that are cut in the <111> plane can be recognized by the triangular pits that will etch in their surface, or the triangular pieces of silicon that result when the wafers are dropped.

If we consider the same set of indices, and take a plane that intersects the x axis at $x = 1$, but never intersects the y or z axes, as shown in Figure 4-2, the Miller indices of the plane are found as follows:

$$x \text{ index} = \frac{1}{x \text{ point of intersection}} = \frac{1}{1} = 1$$

$$y \text{ index} = \frac{1}{y \text{ point of intersection}} = \frac{1}{\infty} = 0$$

$$z \text{ index} = \frac{1}{z \text{ point of intersection}} = \frac{1}{\infty} = 0$$

(For mathematical reasons, a plane that does not intersect an axis is considered to intersect the axis at infinity (∞).)

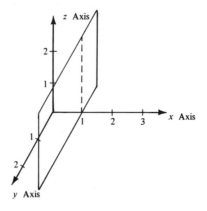

Figure 4-2 A <100> crystal plane.

Thus this is the <100> plane. The <100> plane intersects the crystal axes so that rectangles are formed, therefore <100> crystals form etch pits with square corners, and a <100> wafer will break into rectangular pieces of silicon if it is shattered.

Silicon with either a <111> or <100> crystal orientation is used for almost all device fabrication. These two orientations satisfy almost all requirements, so use of other orientations is generally not required.

4-2 DOPING OF CRYSTALS DURING GROWTH

Following the growth of crystals using either the float zone or the Czochralski method, the concentration of dopant along the crystal depends on both the material used as dopant, and on its original concentration. A term called the "distribution coefficient", k, determines the ratio of the concentration of the dopant in the solid to the concentration of dopant in the liquid.

$$k = \frac{C_s}{C_l} = \frac{\text{Concentration of dopant in the solid phase}}{\text{Concentration of dopant in the liquid phase}} \quad (4\text{-}4)$$

The distribution coefficient of various n-type and p-type dopants is given in Table 4-1.

In Czochralski crystal growth, the gradual freezing of the crystal produces an ever-increasing concentration of dopant in the melt, since the ratio of the concentration of dopant in the solid to that of the

TABLE 4-1: Distribution Coefficients for Common Dopants in Silicon

Dopant	Distribution Coefficient k	Type of Dopant
Phosphorous (P)	.32	n-type
Arsenic (As)	.27	n-type
Antimony (Sb)	.02	n-type
Boron (B)	.72	p-type
Aluminum (Al)	1.8×10^{-3}	p-type
Gallium (Ga)	9.2×10^{-3}	p-type
Indium (In)	3.6×10^{-4}	p-type

liquid is less than one. A typical graph of the concentration as a function of distance along the crystal for $k = .04$ is shown in Figure 4-3.

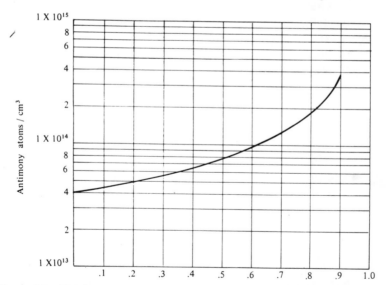

Figure 4-3   The dopant concentration along a Czochralski crystal for an initial dopant concentration of 10^{15} antimony atoms/cm³ (k = .04).

The situation is slightly different in float-zone silicon, but the impurities are swept along in the molten zone as it proceeds from one end of the crystal to the other, resulting in a similar distribution of impurities.

4-3 CRYSTAL DEFECTS

A variety of crystal defects can be present in grown ingots of silicon and other semiconductors. These defects may impact the ability to successfully process wafers from these crystals, or the yield on devices made with these wafers. Two types of defect often encountered are:

1. *Crystal dislocation.* These are localized imperfections in the crystal structure caused by plastic deformation from uneven heating or cooling, or other problems.
2. *Planar slip.* A type of plastic deformation visible because one part of the crystal bar sheared with respect to another.

Both of these common defects, as well as other less common problems can be decorated using a preferential etch. Such an etch attacks silicon along defect boundaries, revealing the nature and extent of the defect.

Crystal defects have been essentially eliminated from silicon grown by using either the float zone or the Czochralski method. However, defects are often inadvertently introduced during subsequent processing, so their study is of continuing importance. Dislocations, slip, and other defects are usually introduced if wafers are improperly heated or cooled during any of the high-temperature processing steps.

REVIEW EXERCISES: WAFER PREPARATION II

1. a. Determine the Miller indices of a plane intersecting a set of axes at
$x = \frac{1}{2}, y = 1, z = \infty$.

b. Sketch the plane in the axes provided.

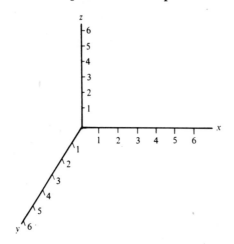

2. What value of k will result in a flat profile for the dopant?

3. What p-type dopant of Table 4-1 will yield the flattest impurity profile during crystal growth?

4. What are the two most common crystal orientations utilized for silicon processing?

5. Describe two types of crystal defects found in wafers cut from grown ingots of silicon.

5

Epitaxial Deposition I

5-1 INTRODUCTION

Epitaxial deposition is the deposition of a single crystal layer on a substrate (often, but not always, of the same composition as the deposited layer), such that the crystal structure of the layer is an extension of the crystal structure of the substrate. One use of epitaxial deposition in semiconductor processing is in the fabrication of light-emitting diodes. The carefully tailored materials and doping profile that result in the generation of light are usually obtained by using epitaxial techniques in conjunction with high-temperature diffusions. A more frequent use of epitaxial deposition is in the production of discrete devices and bipolar integrated circuits using silicon.

In the fabrication of silicon diodes and transistors, devices with higher switching speed, breakdown voltage, or current-handling capability can be obtained using epitaxial deposition. In diode fabrication, a heavily doped silicon substrate used as the starting material will result in low resistance to the flow of current. The high doping level results in a low reverse junction breakdown, though, so an epitaxial layer of lightly doped silicon of the same conductivity type is deposited on the substrate for the actual fabrication of the junction. The cross section of a typical diode fabricated in this manner is shown in Figure 5-1.

A transistor may be fabricated in an analogous manner, using an epitaxial layer for the lightly doped collector region, and diffusing in a base and an emitter. Figure 5-2a shows a transistor fabricated in this manner. The epitaxial layer may also be of the opposite conduc-

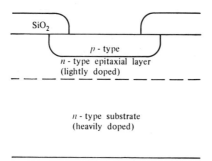

Figure 5-1   Cross section of a diode fabricated using epitaxial silicon.

tivity type, in which case the epitaxial layer serves as the base of the transistor and the emitter is added during a subsequent high-temperature diffusion. A cross section of a device fabricated in this manner is shown in Figure 5-2b.

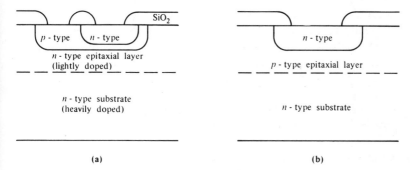

Figure 5-2   Cross sections of transistors fabricated using epitaxial silicon.

In the fabrication of bipolar integrated circuits, a lightly doped silicon substrate of one conductivity type is used as starting material, and a lightly doped epitaxial layer of the opposite conductivity type is deposited on it. (In most cases, a high concentration of the same type dopant used in the epitaxial layer is diffused into regions of the substrate prior to epitaxial deposition, to provide a low-resistance path to the active region of devices.) The substrate helps provide electrical isolation between devices in adjacent pockets when the circuit is in operation. A cross section of a transistor in a typical bipolar integrated circuit is shown in Figure 5-3.

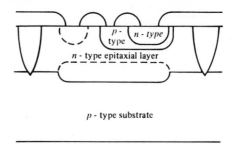

p - type substrate

Figure 5-3   Cross section of a bipolar integrated circuit.

5-2   THEORY

Introduction

Two conditions must be met before epitaxial deposition can occur. First, there must be available sites for the depositing atoms to lose their extra energy and become part of the existing crystal structure. Such sites are called nucleation sites. The availability of suitable nucleation sites greatly influences both the speed with which film growth begins, and the steady-state film growth rate. Second, the atoms to be deposited must reach the substrate and find a lattice site to settle into. These two conditions are separated below to facilitate discussion, but in reality the arrival of atoms at a surface and the availability of nucleation sites cannot be separated. (For the remainder of Chapters 5 and 6, the epitaxial deposition of silicon will be discussed.)

Preparation of Nucleation Sites

Nucleation sites may be prepared prior to placing the substrates in an epitaxial reactor, or may be induced once substrates are in the reactor. In either case, the success with which nucleation sites are prepared is a strong function of the crystal orientation of the substrate. In almost all silicon epitaxial growth, the substrate is 3°–7° off a major axis to allow the easy preparation of nucleation sites by exposing the edges of successive layers of the crystal. Figure 5-4 shows the successive layers of the crystal which are exposed as a result of the substrate crystal orientation being 3° to 7° off a major axis. Figure 5-5 shows the effect of substrate orientation on the deposition rate of silicon.

The use of an etching technique to increase the number of nucleation sites prior to placing the substrate in the reactor must of

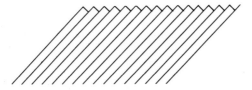

Figure 5-4   The effect of substrate misorientation on the exposure of nucleation sites.

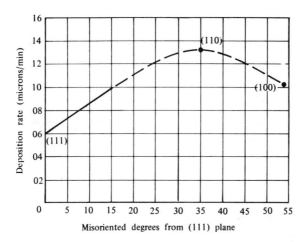

Figure 5-5   Substrate misorientation vs. deposition rate.

necessity use a silicon etchant. Mixtures of hydrofluoric and nitric acid are common silicon etches, and have been used for this purpose. However, the most widely used method of obtaining suitable nucleation sites is to introduce HCl gas just prior to the beginning of actual epitaxial deposition. The HCl etches the top layer of silicon from the substrate, thus removing any crystal defects that may be present at the surface of the wafer. Removal of this layer leaves a surface ready for a subsequent deposition. The etch rate of silicon as a function of HCl concentration in a horizontal reactor is shown in Figure 5-6. The etch rate of silicon is nearly linear for HCl concentrations of 1–4% in hydrogen, so percentages in this range are often used for etching. However, if the fraction of HCl is too high, a pitted substrate surface results. The maximum allowable fraction of HCl in hydrogen at any temperature is shown in Figure 5-7. A total thickness of .25 to 1.0 microns is usually removed from the substrate prior to actual deposition.

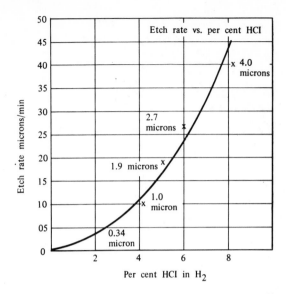

Figure 5-6 Etch rate vs. HCl concentration in hydrogen in a horizontal reactor.

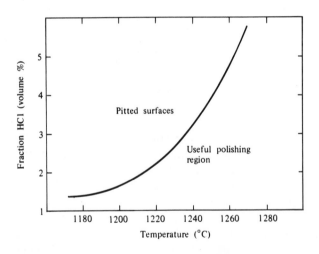

Figure 5-7 Allowable fraction of HCl in H_2 vs. temperature.

Deposition

Vacuum Deposition Under carefully controlled conditions, it has been possible to obtain epitaxial silicon using sputtering or evaporation techniques. However, the low deposition rates and the difficulty of obtaining a high quality crystal structure have prevented these techniques from being used commercially.

Vapor Growth   Epitaxial silicon has been successfully deposited from silicon tetrachloride ($SiCl_4$), Silane (SiH_4), trichlorosilane ($SiHCl_3$), dichlorosilane (SiH_2Cl_2), and other compounds. However, since the first two reactants are used for most industrial epitaxial deposition, only their behavior will be covered in this text.

Hydrogen Reduction of Silicon Tetrachloride   Silicon tetrachloride is commercially available with sufficient purity to provide the lightly doped epitaxial layers needed for device fabrication. It is kept in a carefully controlled constant-temperature bath at temperatures near $0°C$ in a liquid state. Hydrogen flows through or over the $SiCl_4$ to obtain the silicon tetrachloride necessary for deposition. The concentration of $SiCl_4$ in the hydrogen is determined by its flow rate and the temperature of the constant-temperature bath. The effect of temperature on the vapor pressure of $SiCl_4$ is shown in Figure 5-8.

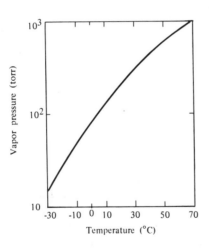

Figure 5-8   Vapor pressure of $SiCl_4$ vs. temperature.

Deposition temperatures of $1150°–1300°C$ are normally used with $SiCl_4$ to obtain a good single crystal layer. This relatively high temperature may cause significant additional diffusion of already present doped regions if care is not taken. The reaction commonly recognized as the one resulting in epitaxial growth is

$$SiCl_4 \ (g) + 2H_2 \ (g) \rightarrow Si \ (s) + 4HCl \ (g). \tag{5-1}$$

However, if excessive $SiCl_4$ is introduced, a competing reaction removes silicon from the substrate:

$$Si \ (s) + SiCl_4 \ (g) \rightarrow 2SiCl_2 \ (g) \tag{5-2}$$

The net result of these reactions is shown in Figure 5-9.

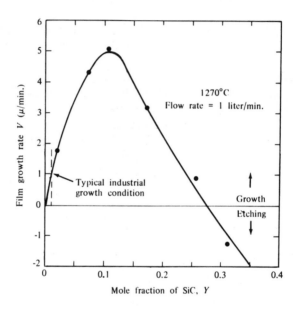

Figure 5-9 Growth rate of silicon vs. mole fraction of $SiCl_4$.

Pyrolysis of Silane Silane is a gas that spontaneously ignites when it comes in contact with air, but is often stored in tanks diluted by hydrogen. The silane or silane/hydrogen mixtures are injected directly into the reactor where the following reaction takes place:

$$SiH_4 \ (g) \rightarrow Si \ (s) + 2H_2 \ (g) \tag{5-3}$$

The growth rate as a function of temperature for this reaction is shown in Figure 5-10. Deposition of epitaxial silicon using silane is usually performed in the 1000°–1100°C range. This deposition temperature range results in less diffusion of previously present heavily doped regions

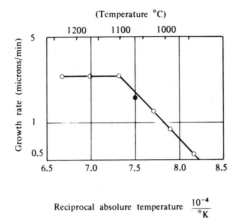

Figure 5-10   Growth rate of silicon vs. temperature for SiH$_4$.

than the use of the higher deposition temperatures required for SiCl$_4$ deposition.

5-3   GROWTH OF AN EPITAXIAL LAYER

The deposition sequence normally followed in an epitaxial process is given below.

1. *Substrate Clean.* The substrates receive a solvent degreasing operation (which may include a physical scrubbing) followed by a series of acid cleaning steps (H$_2$SO$_4$, HNO$_3$: HCl, and HF is a common sequence) and a drying operation. This clean is of great importance, since any residual particles may give rise to imperfections in the deposited layer.

2. *Wafer Load.* Following the completion of the cleaning sequence, extreme care must be taken to insure that the front side of the wafer is not subsequently touched. The use of vacuum wands on the backside of the wafers is the recommended procedure. Equal care must be taken to guarantee that the cleaned substrates never leave regions bathed with filtered air from laminar flow hoods. While placing the substrates on the wafer holder or susceptor, proper precautions will insure that no particles from the susceptor are transferred to the fronts of the substrates.

3. *Heat-up.* Once the epitaxial system has been sealed, a flow of nitrogen purges any residual gases from the system. Following the purge, the reactor heating system is turned on and heat-up begins. Until a temperature of approximately 500°C is reached, nitrogen may be used as the gas flowing through the system. However, since nitrogen begins to etch silicon at elevated temperatures, hydrogen is used to replace the nitrogen at higher temperatures.

4. *HCl Etch.* Once the heat-up cycle is completed and the temperature has been verified using an optical pyrometer or other means, removal of a thin region of damaged silicon at the surface of the wafer using an HCl etch (as previously described) takes place. The amount of silicon removed is carefully controlled to guarantee that the characteristics of the devices being fabricated are not adversely affected.

5. *Deposition.* The deposition step results in an epitaxial layer with the desired thickness and resistivity. Thickness control is obtained by depositing the layer using growth conditions that minimize the error caused by the slight differences encountered in every run. The desired doping concentration is obtained by adding small concentrations of a dopant gas to the main gas flow. The epitaxial layer doping concentration as a function of the dopant to silicon ratio is shown for phosphorus and boron in Figures 5-11 and 5-12.

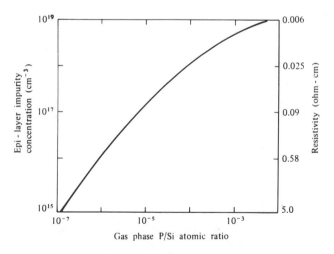

Figure 5-11   Phosphine in gas phase vs. phosphorus in epitaxial layer.

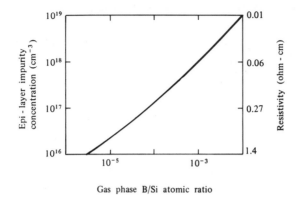

Gas phase B/Si atomic ratio

Figure 5-12 Diborane in gas phase vs. boron in epitaxial layer.

6. *Cool-down.* Following the completion of the growth phase of the process, the temperature is reduced while maintaining the hydrogen gas flow. At approximately 500°C, the gas is switched from hydrogen to nitrogen, and the remainder of the cool-down cycle is completed.

7. *Unloading.* The same degree of care must be taken in the unloading of the silicon wafers as was taken in their loading. The best procedure is to immediately oxidize these wafers to protect their surfaces from possible contamination.

REVIEW EXERCISES: EPITAXIAL DEPOSITION I

1. Must an epitaxial layer be deposited on a substrate of the same composition?
2. Using Figure 5-5, determine the misalignment angle that will result in the highest deposition rate.
3. What is the maximum percentage of HCl that could be used at 1,250°C without resulting in a pitted surface?
4. a. Determine the mole fraction of SiCl$_4$ resulting in a maximum film growth rate?
   b. Why isn't epitaxial silicon grown under these conditions?
5. What two conditions must be met before epitaxial deposition can occur?
6. How are nucleation sites created in a silicon wafer?

7. Describe two disadvantages of vacuum epitaxial deposition.
8. Write the reaction equations and give a brief explanation for the two most widely industrial epitaxial deposition techniques.
9. Determine the thickness of the resulting epitaxial layer using silane at 1,050°C for 5 minutes.

REFERENCES

1. S. K. Tung, "The Effects of Substrate Orientation on Epitaxial Growth." *J. Electrochem. Soc. 112* (April 1965), pp. 436–438.
2. K. E. Bean and P. S. Gleim, "Vapor Etching Prior to Epitaxial Deposition of Silicon." Paper presented at the Fall meeting of the ECS, 1963.
3. G. A. Lang and T. Stavish, "Chemical Polishing of Silicon with Anhydrous Hydrogen Chloride," *R.C.A. Review 24* (December 1963), pp. 488–498.
4. K. J. Miller and N. J. Frieco, "Epitaxial *P*-type Germanium and Silicon Films by the Hydrogen Reduction of GeBr$_4$, SiBr$_4$ and BBr$_3$." *J. Electrochem. Soc. 110* (December 1963), pp. 1252–1256.
5. H. C. Theuren, "Epitaxial Silicon Films by the Hydrogen Reduction of SiCl$_4$." *J. Electrochem. Soc. 108* (July 1961), pp. 649–653.
6. B. A. Joyce and R. R. Bradley, "Epitaxial Growth of Silicon from the Pyrolysis of Monosilane on Silicon Substrates," *J. Electrochem. Soc. 110* (December 1963), pp. 1235–1240.
7. R. M. Warner, Jr., ed., Motorola, Inc. Semiconductor Products Division, *Integrated Circuits* (New York: McGraw-Hill), 1965.

6

Epitaxial Deposition II

6-1 INTRODUCTION

Equipment for epitaxial deposition must meet a severe set of requirements. Since silicon is deposited on substrates from the gaseous state, the reaction chamber must be free from leakage over an extended temperature range. This requirement is usually met by using a quartz reaction chamber. The flow of gases into the reaction chamber is controlled and monitored at all times. The wafers in the reaction chamber rest on a holder called a susceptor. In addition to supporting the wafers, the susceptor serves as the local source of heat if the reactor is heated using a radio frequency (RF) generator. Radio frequency (or induction) heated reactors work by inductively coupling an electromagnetic field into the susceptor. Susceptors usually have a graphite (carbon) body with a thin coating of silicon carbide over the outer surface. The silicon carbide coating prevents the contamination of device wafers with carbon. The wafers resting on the susceptor are heated because they are physically in contact with it. In an epitaxial process, the deposition proceeds most rapidly on the hottest surfaces in the chamber, which are the front surfaces of the wafers and the susceptor. Deposition on the cooler chamber walls proceeds at a considerably lower rate.

An alternate method of heating wafers for epitaxial deposition is to use ultraviolet energy. Special light bulbs that emit large amounts of ultraviolet radiation heat the substrates through a transparent quartz window. The ultraviolet radiation directly heats the wafers and the susceptor that supports them. The desired temperature is maintained using a temperature sensor and a controller that corrects for deviations from a predetermined value.

Three general configurations of epitaxial reactors are currently being used in industrial applications. These systems are known as:

1. Vertical systems   (Figure 6-1)
2. Horizontal systems   (Figure 6-2)
3. Barrel or cylinder systems   (Figure 6-3)

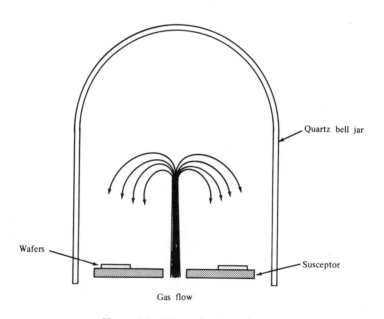

Figure 6-1   Vertical epitaxial reactor.

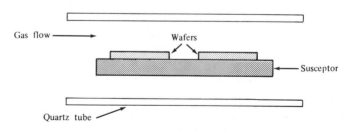

Figure 6-2   Horizontal epitaxial reactor.

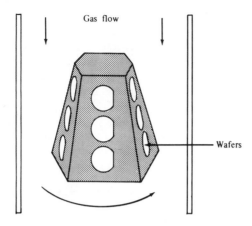

Figure 6-3 Barrel or cylinder reactor.

In a vertical epitaxial deposition system, the gases enter the chamber near the bottom and swirl in the chamber to react at the surface of the susceptor. The susceptor rotates, helping to maintain a uniform temperature on the susceptor and distributing the gases throughout the reaction chamber. The gases exhaust from the bottom of the chamber.

In a horizontal system, the gases enter the chamber at one end and are exhausted at the other. Care is taken to guarantee that the required temperature profile is maintained. The susceptor is often inclined a few degrees to produce a gas-flow profile that introduces an unused gas stream at every position on the susceptor as shown in Figure 6-2.

A barrel system combines features of both the horizontal and the vertical reactor. In a barrel system, wafers are placed on the faces of a rotating susceptor. Each face of the susceptor is thus the equivalent of one susceptor in a horizontal system. The main advantage this configuration offers is the ability to deposit epitaxy on a large number of wafers simultaneously.

A fourth type of reactor is in design, though additional work is necessary to make it commercially available. This type of reactor has been called a carousel, since wafers are held in a carousel-like wafer holder (Figure 6-4 page 46).

6-2 EVALUATION OF EPITAXIAL LAYERS

The three important parameters in epitaxial deposition are:

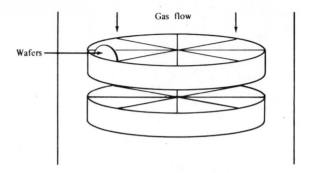

Figure 6-4   The gases pass through the tiers of rotating carousels.

1. The thickness of the deposited layer, and its variation across the wafer and the run.
2. The concentration of desired impurities and its variation across the wafer and the run.
3. The density and distribution of crystal defects in the deposited layer.

All of these parameters must be within certain bounds for the epitaxial layer to successfully perform its intended purpose.

Epitaxial layer thickness can be measured in a number of ways. Three of these ways are outlined below.

1. *Angle lap and stain (or groove and stain).* The epitaxial layer is deposited on a substrate of the opposite conductivity type (*p* on *n* or *n* on *p*). This epi-layer is then grooved or polished beyond its boundary with the substrate. A staining solution is applied to the exposed junction to delineate the junction by darkening either the *p*-type or the *n*-type surface. Monochromatic light is then used with a glass cover to generate interference fringes which determine the junction depth. The junction depth is related to the number of fringes by the formula:

$$d = \frac{n\,\lambda}{2}$$

where λ = the wavelength of the monochromatic light
         n = the number of interference fringes
         d = the junction depth

2. *Etch pit depth.* The presence of defects at the interface between the substrate and the epitaxial layer generates defects that propagate to the surface of the wafer along the crystal planes. A preferential etch is used to delineate the etch pits. The thickness of the epitaxial layer is geometrically related to the length of the sides of the exposed etch pits. For <111> silicon shown in Figure 6-5, $d = .816\, a$.

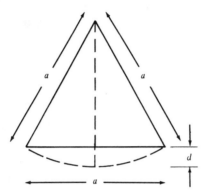

Figure 6-5   Etch pit determination of epitaxial layer thickness.

3. *Infrared interference.* The boundary between the substrate and the epitaxial layer represents an interface that reflects light rays of the correct length. By determining the wavelength of light that produces certain interference responses, the thickness can be determined as shown in Figure 6-6.

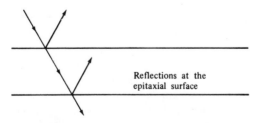

Reflections at the epitaxial surface

Figure 6-6   Use of infrared interference to determine epitaxial layer thickness.

The impurity concentration can be determined using various techniques. Unfortunately, they are all rather time-consuming and tedious.

1. *Sectioning.* The layers of silicon are removed from the surface of the wafer using anodic oxidation or etching, and the sheet resistance of the newly exposed surface is determined. This data can be mathematically manipulated to give the impurity concentration.

2. *Reverse bias C-V technique.* A layer of metal is deposited on the surface of the layer to form a Schottky barrier diode. This diode is reverse biased, and the capacitance as a function of reverse-biased voltage (C-V) is determined. This information is then mathematically manipulated to give the impurity concentration.

3. *Lap and spreading probe.* The profile to be measured is exposed using a lapping technique, and a fine probe is used to determine the resistivity of the material at positions along the exposed surface of the profile as shown in Figure 6-7.

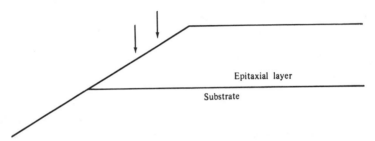

Epitaxial layer

Substrate

Figure 6-7 Lap and spreading probe technique for determining impurity concentration.

In a fashion similar to that described above, the data are reduced, resulting in a plot of the dopant profile as a function of depth into the wafer.

The quality of the epitaxial layer is measured using an etch technique that preferentially exposes defects in the crystal structure. The number, location, and kind of defect determines the crystal quality of the epitaxial layer, as was discussed in Chapters 3 and 4.

REVIEW EXERCISES: EPITAXIAL DEPOSITION II

1. Name and discuss two methods of heating wafers in an epitaxial reactor.
2. Why is epitaxial deposition done using a cold-wall reactor?
3. What are three important parameters in epitaxial deposition?
4. The wavelength of light used to determine a junction depth is $.3\mu$. If 8 fringes are present, what is the epitaxial layer thickness?
5. List and describe two methods of determining the thickness of an epitaxial layer.
6. Determine the thickness for (111) silicon of an epitaxial layer in which the sides of the etch pits are $1.838\ \mu m$.

7

Oxidation I

7-1 INTRODUCTION

The ability to grow a chemically stable protective layer of silicon dioxide (SiO_2) on silicon, makes silicon the most widely used semiconductor. This protective layer is grown in atmospheres containing either oxygen (O_2) or water vapor (H_2O) at temperatures in the range of 900°–1300°C. The process of oxidation can be investigated by considering a surface of silicon with an already existing layer of silicon dioxide on it. (Figure 7-1).

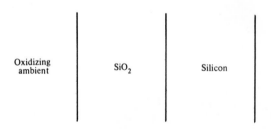

| Oxidizing ambient | SiO_2 | Silicon |

Figure 7-1 Silicon with a layer of silicon dioxide on its surface.

Except for the first few moments, a silicon slice will nearly always have a SiO_2 layer on it, so this assumption is a valid one. Oxidation takes place when either oxygen or water vapor reacts with the silicon as shown in the chemical equations,

$$Si + O_2 \rightarrow SiO_2 \qquad (7\text{-}1)$$

$$Si + 2H_2O \rightarrow SiO_2 + 2H_2 \qquad (7\text{-}2)$$

For the silicon and the oxidizing species to react, one of the following must take place:

1. The oxidizing species must diffuse through the layer of SiO_2 to reach the silicon–SiO_2 interface where the reaction takes place (Figure 7-2a).
2. The silicon must diffuse through the layer of SiO_2 to the interface between the silicon dioxide layer and the oxidizing atmosphere where the reaction takes place (Figure 7-2b).
3. The two active species meet somewhere in the SiO_2 layer where the reaction takes place (Figure 7-2c).

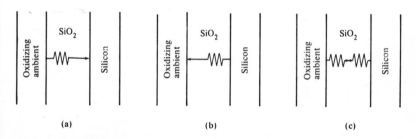

Figure 7-2 Potential reaction mechanisms in the oxidation of silicon; (a) oxidizing species diffuses through the SiO_2 layer interface; (b) silicon diffuses through the SiO_2 surface; (c) the oxidizing species and the silicon meet in the SiO_2 layer.

Experiments have disclosed that in the thermal oxidation of silicon, the first process, the diffusion of the oxidizing species (either O_2 or H_2O) through the existing layer of SiO_2, occurs.

The assumption is that we can oxidize through the SiO_2 layer on the surface of the silicon slice. It is necessary to make this assumption since virtually all silicon wafers, once created, will always have a surface layer of SiO_2.

Experiments verify that we *can* oxidize through this initial thin layer to create thicker oxides and partially consume the silicon.

7-2 EQUIPMENT FOR THERMAL OXIDATION

Thermal oxidations are performed in furnaces in which the temperature is carefully controlled (limits of $\pm\frac{1}{2}$ °C are typical). Furnaces generally contain three or four sets of coils, each with its own set of controls and quartzware. The coils are heated electrically, and the current is adjusted and controlled to provide the required constant temperature. A quartz tube (or occasionally a tube made from another material such as silicon or silicon carbide) rests inside the coils, providing an envelope around the wafers in which the atmosphere can be controlled. A cross section of a typical oxidation furnace is shown in Figure 7-3.

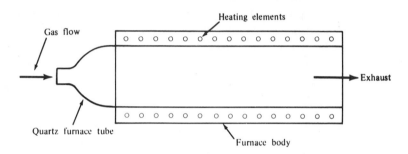

Figure 7-3 Cross section of an oxidation furnace.

7-3 THE OXIDATION PROCESS

The thermal oxidation of silicon must be preceded by a cleaning sequence designed to remove all contamination. Particular care must be taken during this handling to guarantee that the wafers do not contact any source of contamination, particularly inadvertent contact with a person. (Humans are a potential source of sodium, the element most often responsible for the failure of devices due to surface leakage.) The cleaned wafers are dried, loaded into a quartz wafer holder called a "boat," and are ready for oxidation.

Thermal oxidation using dry oxygen involves controlling the flow of oxygen into the quartz tube to guarantee that an excess of oxygen is available for the silicon. A source of high-purity oxygen makes sure that no unwanted impurities become incorporated in the layer of oxide as it grows. Oxygen or an oxygen/nitrogen mixture is used for growing the layer of oxide. The use of nitrogen decreases the total cost of running the oxidation process, as it is less expensive than oxygen.

Three methods of introducing water vapor are commonly used when water is the oxidizing species. Water may be placed in a container called a "bubbler" and maintained at a constant temperature below its boiling point (100°C). A bubbler is shown in Figure 7-4 with a heating mantle that keeps it at a predetermined temperature.

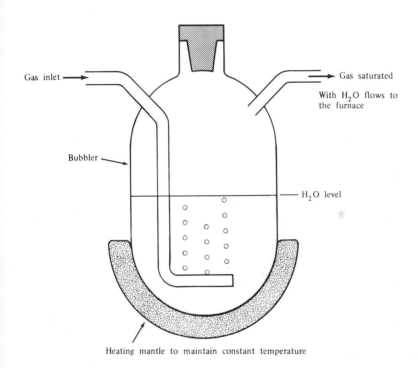

Figure 7-4 A bubbler for a wet oxidation system.

Gas enters the inlet side of the bubbler, becomes saturated with water vapor as it rises through the water, and exits through the outlet into the furnace. The distance from the outlet to the quartz oxidation tube must be short enough to prevent condensation by cooling, or must be heated using an auxiliary method. Nitrogen or oxygen may be used in the carrier gas, with equivalent oxide thickness being grown regardless of which gas is used. Maintenance of a constant temperature is important because the vapor pressure of water varies with temperature as shown in Figure 7-5.

The temperature of the water is also generally kept a couple of degrees below its boiling point to afford better control of the vapor

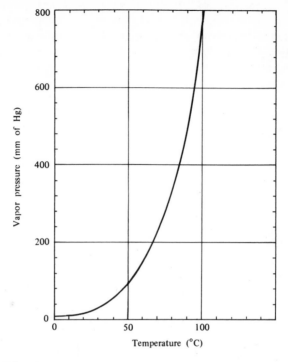

Figure 7-5 The vapor pressure of water as a function of temperature.

pressure. If the temperature is too close to 100°C, small temperature variations produce large changes in the vapor pressure. Bubblers are simple to use and quite reproducible, but have two disadvantages associated with the fact that they must be refilled when the water level falls too low.

1. Improper handling of container, etc., can result in the contamination of the water prior to or during filling.
2. The bubbler cannot be filled during a cycle unless heated water is used. If cool water is added, the vapor pressure of the water will decrease when the new water is introduced.

A second method of obtaining water vapor is the introduction and subsequent combustion of a hydrogen/oxygen gas mixture. Such systems are often called "burnt hydrogen," or "torch" systems. Water vapor is produced when proper amounts of hydrogen and oxygen are introduced into the inlet end of the tube and allowed to react. A quartz injector with a specially shaped tip is used to guarantee proper combustion of the mixture. The heat produced at the inlet end of the oxidation furnace often makes it necessary to reprofile the furnace under operating conditions to guarantee a uniform temperature profile along the entire tube. The installation of these units necessitates the availability of both oxygen and pure hydrogen. Because the potential for an explosion exists if excess hydrogen is introduced, an explosion preventive system is a necessity on installed units.

A third method used for wet oxidation is the use of a system often called a "flash" system. Such a system is shown in Figure 7-6. Water drips continuously onto the heated bottom surface of the units, evaporating rapidly once it touches the surface. A carrier gas, either oxygen or nitrogen, flows over the evaporating water, carrying the water vapor into the furnace. This type of oxidation unit needs only a steady supply of pure water to function.

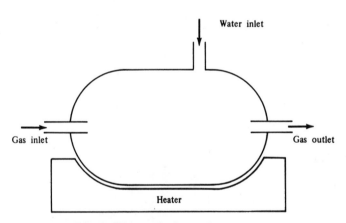

Figure 7-6 A "flash" oxidation system.

7-4 OXIDE EVALUATION

The two important characteristics of a layer of SiO_2 are its thickness and its quality. The thickness can be accurately predicted from the oxidation sequence, but it is often necessary to verify results. The thin, uniform layers of SiO_2 useful in most processing steps, produce colors when observed perpendicularly in white light. A chart of oxide colors and associated oxide thicknesses is given in Table 7-1. As the thickness of oxide increases from a bare wafer to very thick layers, the colors observed with these thicknesses repeat themselves. To determine the

TABLE 7-1: Color Chart for Thermally Grown SiO$_2$ Films

(OBSERVED PERPENDICULARLY UNDER DAYLIGHT FLUORESCENT LIGHTING)

Film Thickness (microns)	Order (5450 A)	Color and Comments	Film Thickness (microns)	Order (5450 A)	Color and Comments
0.050		tan	0.365	II	yellow-green
0.075		brown	0.375		green-yellow
			0.390		yellow
0.100		dark violet to red violet	0.412		light orange
0.125		royal blue	0.426		carnation pink
0.150		light blue to metallic blue	0.443		violet-red
0.175		metallic to very light yellow-green	0.465		red-violet
	I		0.476		violet
0.200		light gold or yellow—slightly metallic	0.480		blue-violet
0.225		gold with slight yellow orange	0.493		blue
0.250		orange to melon			
0.275		red-violet	0.502		blue-green
			0.520		green (broad)
0.300		blue to violet-blue	0.540	III	yellow-green
0.310		blue	0.560		green-yellow
0.325		blue to blue-green	0.574		yellow to "yellowish" *
0.345		light green	0.585		light orange or yellow to pink borderline
0.350		green to yellow-green			

TABLE 7-1 (*cont.*)

Thickness	Group	Color
0.60		carnation pink
0.63		violet-red
0.68		"bluish" **
0.72	IV	blue-green to green (quite broad)
0.77		"yellowish"
0.80		orange (rather broad for orange)
0.82		salmon
0.85		dull, light red-violet
0.86		violet
0.87		blue-violet
0.89		blue
0.92	V	blue-green
0.95		dull yellow-green
0.97		yellow to "yellowish"
0.99		orange
1.00		carnation pink
1.02		violet-red
1.05		red-violet
1.06		violet
1.07		blue-violet
1.10		green
1.11		yellow-green
1.12	VI	green
1.18		violet
1.19		red-violet
1.21		violet-red
1.24		carnation pink to salmon
1.25		orange
1.28		"yellowish"
1.32	VII	sky blue to green-blue
1.40		orange
1.45		violet
1.46		blue-violet
1.50	VIII	blue
1.54		dull yellow-green

* Not yellow but is in the position where yellow is to be expected; at times it appears to be light creamy grey or metallic.

** Not blue but borderline between violet and blue-green; it appears more like a mixture between violet-red and blue-green and overall looks greyish.

NOTE: Above chart may also be used for Vapox, Sputtox, Phosphox and Borox dielectric films. For silicon nitride films, multiply film thickness by 0.75.
SOURCE: *IBM J. Res. Dev.*, 8, 43 (1964).

thickness of a layer of oxide from its color, you must also know all the colors preceding it on the color chart. This requirement is easily met by etching a taper in the layer of oxide by immersing it in hydrofluoric acid and slowly withdrawing it (Figure 7-7). The thickness can also be measured electrically by measuring the capacitance it produces between two conductive plates of known area, or by etching a step in the layer of oxide and using optical interference or physical techniques. An instrument often used to determine the thickness of a step is called a surface profilometer. A stylus with an electronically amplified output is drawn over the surface to be measured, and an ink trace of the resulting profile shows the height of any steps in the surface. Very accurate determinations of oxide thicknesses within certain ranges can be made using an optical instrument called an ellipometer. However, the degree of accuracy that is attainable using this instrument is necessary in only a few instances.

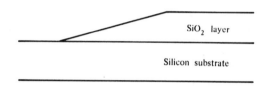

Figure 7-7   Tapered oxide thickness on a silicon wafer.

The dielectric quality of a layer of SiO_2 is usually determined by two parameters:

1. The breakdown strength of a layer of SiO_2.
2. The amount of contamination present in the SiO_2 layer that drifts under an applied voltage.

The first parameter is measured by placing a voltage between two conducting plates across a known thickness of SiO_2, and increasing the voltage until the current between the two plates increases significantly. The breakdown value is measured for a predetermined number of areas on a wafer, with the quality of the oxide determined by the distribution of breakdown voltage. For silicon dioxide, 600 V / um. is considered an acceptable dielectric strength. The amount of mobile contamination (generally sodium) is measured using a technique called Capacitance-Voltage or C-V. This analysis technique uses a thin layer of SiO_2 with conducting plates on either side. The capacitance of the structure is measured after it has been biased first negative at one terminal with respect to the other, and then positive at one terminal with respect to the other at an elevated temperature. The difference between the C-V curves at these extremes is related to the amount of mobile contamination.

7-5 RECENT ADVANCES IN OXIDATION TECHNOLOGY

Recent studies have indicated that the measured dielectric quality of a layer of SiO_2 can be increased while the apparent amount of mobile contamination is decreased if a chlorine-carrying compound such as HCl or TCE is injected into the oxidation tube during the growth of the oxide layer. The amount of chlorine-carrying species must be controlled within certain bounds to obtain the benefits. It has been hypothesized that the chlorine ion accumulates near the $Si-SiO_2$ interface where it is able to combine with any mobile contamination it encounters, rendering it immobile. During an oxidation cycle, it may similarly tie up any contamination it encounters.

REVIEW EXERCISES: OXIDATION I

1. Of what material are oxidation tubes usually fabricated?
2. What are the two chemicals used in the oxidation of silicon?
3. In the oxidation of silicon, where does the chemical reaction take place?
4. List and briefly describe three methods of introducing water vapor into an oxidation furnace.
5. What is the purpose of nitrogen during the dry O_2 oxidation process?
6. At what temperature typically is the water in an oxygen bubbler maintained?
7. What is the potential hazard with the "burnt hydrogen" oxidation system?
8. What is the typical breakdown voltage of a layer of SiO_2 two microns thick?
9. What method may be utilized to increase the dielectric quality of a layer of SiO_2?

8

Oxidation II

8-1 OXIDE THICKNESS DETERMINATION

Though both oxygen and water vapor are used to oxidize silicon wafers, oxidations using these two species are not interchangeable. Oxidation using water vapor proceeds at a significantly faster rate than oxidation using oxygen, for the same temperature and time. The different oxidation rates for these two oxidizing species give rise to a different set of applications for each type of oxidation. The oxide thickness resulting from a single oxidation step starting with a bare silicon wafer can be determined using Figures 8-1 and 8-2. Some specific examples of oxide thickness determination are given below:

Example 1: Determine the SiO_2 thickness following a 50 minute, 920°C steam oxidation cycle.

Solution: Using Figure 8-2, find the line labeled 920°C and determine its intersection with the line denoting 50 minutes. From this point of intersection proceed directly across to the left-hand edge of the graph. The thickness indicated is 2300 Å or .23μ.

Example 2: Determine the SiO_2 thickness following a 90 minute, 1000°C dry O_2 oxidation cycle.

Solution: Using Figure 8-1, find the line labeled 1000°C and determine its intersection with the vertical line representing 90 minutes. From this point of

intersection proceed directly across to the left-
hand edge of the graph. The intersection along
this axis is .085μ or 850 Å.

In typical device processing, however, a silicon wafer will see
a sequence of oxidation cycles, with both the temperatures and the
oxidizing species varying. The thickness of an oxide layer following
any series of oxidation steps can be determined as long as a few rules
are followed.

1. Always begin with the present thickness of SiO_2. Determine
how long it takes to grow this thickness using the next-
growth conditions. (If the wafer is bare, time is zero.)
2. To the amount of time determined in step **1**, add the additional oxidation time in the present cycle.
3. Find the oxide thickness resulting from the time in step **2**.
If the cycle was the last oxidation cycle, then you have the
total thickness. If not, use this thickness as the starting point
and go through steps **1**, **2**, and **3** again.

As an example of this technique, consider the following sequence of oxidations:

1. Dry O_2 1,200°C   60 minutes
2. 95°C Steam   900°C   40 minutes
3. 95°C Steam   1,200°C   13 minutes

The oxide thickness following the first cycle can be determined directly
from Figure 8-1. We see from this figure that the thickness of the first
cycle is 2000 Å. Now, using this as the starting thickness, determine
how long it takes to grow the 2000 Å layer using the growth condition
of the second oxidation step. The next oxidation takes place at 900°C in
steam. Find the point at which the 2000 Å line crosses the 900°C
growth line on Figure 8-2. The time to grow 2,000 Å of SiO_2 at 900°C
is 50 minutes. As far as the silicon is concerned, the 2,000 Å of SiO_2
present on the surface of the wafer could have taken 50 minutes to
grow at 900°C. We next add another 40 minutes at 900°C to the
growth time, resulting in the equivalent of 90 minutes at 900°C. From
the figure, 900°C for 90 minutes at these growth conditions results in
3,000 Å of SiO_2.

With 3,000 Å as a starting point, follow this procedure once
again. Determine the point at which the 3,000 Å line intersects the
1,200°C growth curve on Figure 8-2. We see that it would have taken

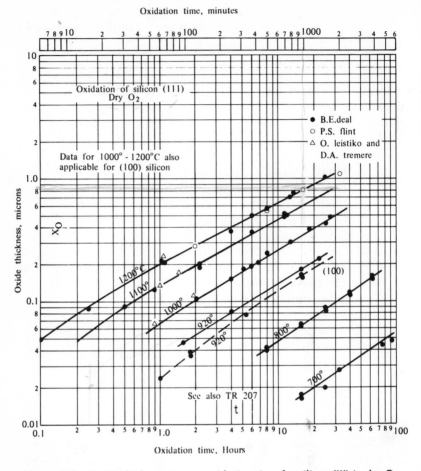

Oxidation time, minutes

Figure 8-1 Oxide thickness versus oxidation time for silicon (III) in dry O_2.

9 minutes to grow 3,000 Å at this growth condition. This 9 minutes is used as the starting time, and adding the additional 13 minutes at 1,200°C, we end up with 22 minutes at 1,200°C in steam. Looking up to the 1,200°C growth curve, we see an intersection at the 5,000 Å line, so the final oxide thickness is 5,000 Å.

8-2 OXIDATION REACTION

An understanding of the physical process that takes place during an oxidation can be obtained by considering the two extreme cases of a reaction. These two cases are known as:

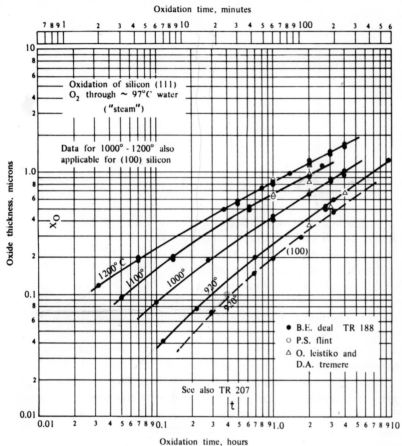

Figure 8-2 Oxide thickness versus oxidation time for silicon (III) in wet O_2.

1. Transport-limited case.
2. Reaction rate-limited case

These two extreme cases can be understood by considering Figure 8-3. Once any amount of oxide has grown on the surface of a silicon wafer, the oxidizing species must move through this oxide layer to reach the silicon. The oxide growth can be limited by:

1. The availability of the oxidizing species at the Si–SiO_2 interface.
2. The ability of the reaction between the oxidizing species and the silicon to take place.

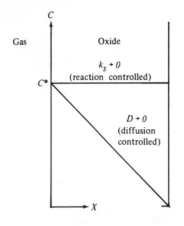

Figure 8-3 Distribution of the oxidizing species in the oxide layer for the two limiting cases of oxidation.

When the oxide layer is sufficiently thick, the oxidizing species cannot diffuse through this layer rapidly enough to keep the reaction going at peak speed. This extreme case is the "transport-limited" or "diffusion-limited" case. The presence of a very thin layer of SiO_2 does not interfere with the diffusion of oxidizing atoms to the Si–SiO_2 interface. These atoms diffuse to the interface until an excess of them is present. The SiO_2 growth rate is limited by the speed with which the silicon can react with the oxidizing atoms.

The different growth rates of reaction rate-limited and transport-limited cases can be seen in Figures 8-1 and 8-2. For short growth times at low temperatures, the slope of the oxide thickness vs. time curve is different than the one for high temperatures and long growth times. The steeper slope corresponds to the reaction rate-limited case and the less steep slope to the transport-limited case.

8-3 REDISTRIBUTION OF DOPANT ATOMS DURING THERMAL OXIDATION

During the course of thermal oxidation, the interface between the silicon layer and the silicon dioxide layer moves through doped regions of the silicon. Dopants present at this moving interface will redistribute themselves depending on their relative solubility in silicon and silicon dioxide. Phosphorus, arsenic, and antimony have greater solubilities in silicon than in silicon dioxide, so these impurities tend to pile-up in front of an advancing Si–SiO_2 interface. Boron on the other hand, has a greater solubility in silicon dioxide, so boron is depleted from the silicon in front of an advancing Si–SiO_2 interface, and accumulates in the newly-grown silicon dioxide layer. The redistribution of phosphorus and boron during thermal oxidation is shown in Figures 8-4 and 8-5.

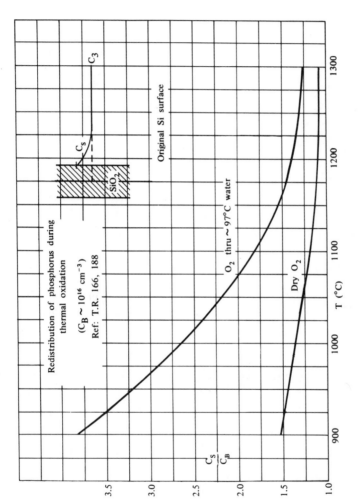

Figure 8-4   Redistribution of phosphorus during thermal oxidation.

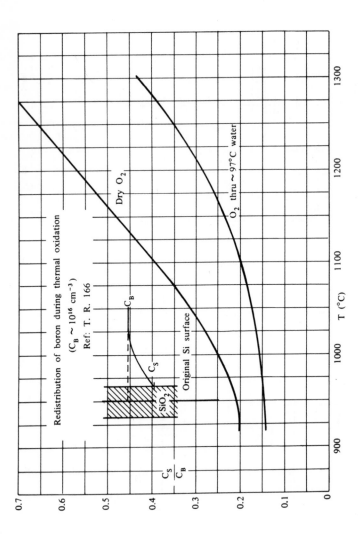

Figure 8-5 Redistribution of boron during thermal oxidation.

8-4 ANODIC OXIDATION

Anodic oxidation is a technique for growing a relatively thin layer of SiO_2 (up to ~600 Å) on a wafer at low temperatures. Using this technique, a positive voltage is applied to a silicon wafer in an electrolytic solution, making it the anode. A cathode is placed in the solution, and a voltage is applied from anode to cathode (Figure 8-6). The voltage from anode to cathode determines the final oxide thickness on the wafer. The oxide layer formed using this technique is of poor quality from an electrical standpoint, but silicon layers of reproducible thickness are easily removed in this manner. In anodic oxidation, the oxidizing species diffuses through the oxide layer to the SiO_2–silicon interface, leaving the profile of any previously diffused dopant unchanged. A combination of anodic oxidation, SiO_2 removal, and four-point probing is often used to determine the profile of a diffusion.

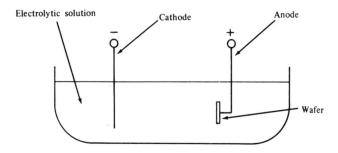

Electrolytic solution   Cathode   Anode   Wafer

Figure 8-6   Apparatus for the anodic oxidation of silicon.

REVIEW EXERCISES: OXIDATION II

1. Determine the thickness of a layer of SiO_2 grown on a bare silicon wafer if the oxidation cycle is:
   a. 1,200°C   Dry O_2 60 minutes
   b. 1,000°C   97°C H_2O 12 minutes

2. If the time a wafer is oxidized is doubled, does the oxide thickness double? If not, why not?

3. Determine the thickness of an SiO_2 layer grown on a bare silicon wafer in the following three sequential steps:
   a. 4.5 hours .   1,100°C   Dry O_2
   b. 6 minutes   1,200°C   97°C H_2O
   c. 12 minutes   1,100°C   97°C H_2O

4. Is silicon or the oxidizing species (either O_2 or H_2O) the mobile species in anodic oxidation?

5. A (111) silicon wafer is oxidized in steam at a temperature above 1,200°C for 30 minutes to obtain an oxide thickness of 1 micron. How much longer will it take to obtain a total oxide thickness of 3 microns at the same temperature and ambient conditions?

6. A (100) silicon wafer is oxidized for 24 minutes in steam at 1,100°C. How much time is needed to grow an additional one micron of oxide in dry O_2 at 1,000°C and what is now the total oxide thickness?

7. Explain why steam oxidation of silicon proceeds at a faster rate than dry oxidation.

8. Explain the difference between transport limited and reaction rate limited oxidation.

9. When boron-doped silicon is oxidized, is the tendency for the boron in the silicon to pile up or be depleted at the silicon-oxide interface? Explain your reasoning.

9

Impurity Introduction and Redistribution I

9-1 THE IDEA OF DIFFUSION

Diffusion is a term used to describe the movement of particles away from regions of high concentration. An example of the process of diffusion can be visualized by thinking of a small quantity of black ink being dropped into a still glass of clear water. Initially, the ink is a dark region in the clear water, but gradually some of the ink moves away from this region, and instead of there being just a dark and a clear region, there is a gradation of colors. After more time passes, the ink spreads out until it is possible to see through it, though some regions are darker than others. Finally, after a very long time, the ink is uniformly distributed in the water. The movement of the ink from the high concentration region (the initial ink drop) to the low concentration region (the rest of the glass of water) is the process of diffusion.

The process of diffusion is also represented in Figure 9-1. We start with Figure 9-1a which shows the distribution of particles at the initial time. As time passes, the distribution moves away from the center in both directions as shown in Figures 9-1b, 9-1c. Finally, after a very long time, the particles are uniformly distributed as shown in Figure 9-1d.

The rate at which the diffusion of particles takes place depends on how fast they are moving. This velocity depends in turn on the temperature, since hotter particles move faster. The "diffusion coefficient" of a material is the term used to describe this dependence on temperature.

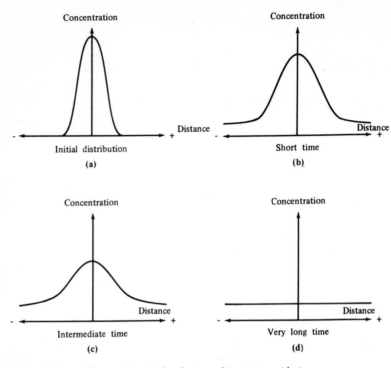

Figure 9-1 Redistribution of impurities with time.

9-2 THE DIFFUSION PROCESS

The process of diffusion is used in semiconductor processing to introduce a controlled amount of chosen impurities into selected regions of
a semiconductor crystal. The diffusion used in processing can be divided into two distinct types.

1. Predeposition: the introduction of a carefully controlled
   amount of desired impurity into the semiconductor.
2. Drive-in: the redistribution of the impurity to obtain the
   final profile in the semiconductor.

We will consider each of these in great detail in the next section.

The Predeposition

During the predeposition, the semiconductor substrates are heated to a
carefully selected and controlled temperature, and an excess of the desired dopant is made available at the surface of the wafer. (We will

discuss how this excess is obtained below.) The materials used as dopants will enter the crystal structure until a maximum concentration called the "solid solubility" is reached. The solid solubility of one material in another depends on the temperature alone. The solid solubility of common dopants in silicon is shown in Figure 9-2.

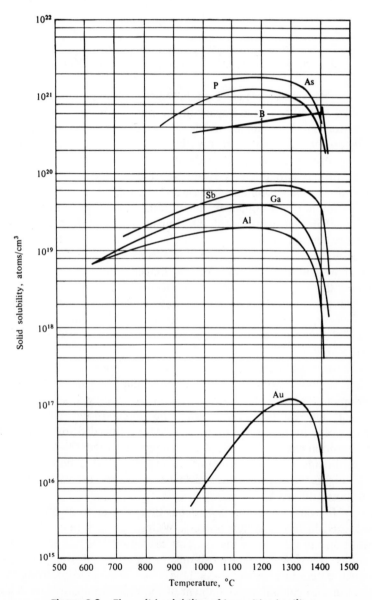

Figure 9-2   The solid solubility of impurities in silicon.

The solid solubility of a dopant in silicon at a given temperature is the maximum amount of that dopant that can be present in the silicon. To obtain controlled conditions during a predeposition, an excess of dopant is present at the surface of the wafer. Having more dopant available outside the silicon than can enter the silicon guarantees that solid solubility will be maintained during the predeposition. Use Figure 9-2 as an example to determine:

1. The solid solubility of phosphorus in silicon at 1,000°C.
2. The solid solubility of boron in silicon at 1,200°C.

The temperature at which the substrate is maintained determines the solid solubility of the dopant in the semiconductor, and hence the concentration of the dopant at the surface of the wafer. However, the predeposition time is the other variable needed to completely characterize a predeposition. The predeposition time determines the exact doping profile as one moves away from the surface of the wafer. The effect of ever-increasing time is shown in Figure 9-3. Figure 9-3a shows the distribution of a dopant in silicon after a small period of time has elapsed. The concentration of the dopant at the surface is the solid solubility determined from Figure 9-2. The dopant concentration quickly decreases as we move away from the surface. Figure 9-3b and Figure 9-3c show the dopant profile for two longer times. The concentration remains the same at the surface of the wafer, but the dopant profile falls off less rapidly as we move away from the surface. Finally, in Figure 9-3d, a very long period of time has elapsed. The dopant profile is now flat throughout the wafer, and at the limit determined by the solid solubility.

In the fabrication of silicon devices, the presence or absence of a layer of silicon dioxide on the surface of the wafer determines where the dopant is allowed to enter the silicon. As long as a silicon dioxide layer of sufficient thickness is used, the predeposition will introduce dopants only in the desired area. The thickness needed for a particular predeposition can be determined experimentally. Figure 9-4a shows the silicon dioxide thickness necessary to mask against a boron diffusion while Figure 9-4b shows the thickness necessary to mask against a phosphorus diffusion.

In wafer fabrication, predepositions are done in furnaces like those used in the oxidation process. Wafers that are ready for a predeposition are cleaned to remove any contamination picked up during previous steps. Most predepositions are performed by placing the cleaned wafers in a quartz wafer holder or "boat" and inserting them into a furnace containing an ambient of the desired dopant. The dopant

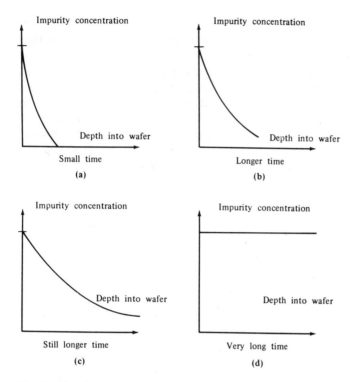

Figure 9-3   Profile of dopant present in a wafer as a function of time for increasing time.

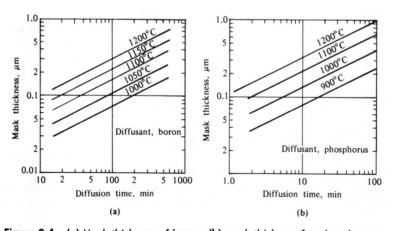

Figure 9-4   (a) Mask thickness of boron; (b) mask thickness for phosphorus.

is carried into the furnace by gas flow from the source end of the quartz tube. Enough dopant is deposited on each wafer to guarantee that the solid solubility limit is reached.

The compounds used as dopant sources for this type of predeposition may be solids, liquids, or gases. Solids are often used in powder form. They are heated, and a carrier gas is passed over them carrying the dopant into the furnace. The powder may be heated in the end of the furnace itself, or a separate furnace called a source furnace may be used. The arrangement of the source end of a predeposition furnace using a powder is shown in Figure 9-5. The carrier gas used during most predeposition of this nature is nitrogen.

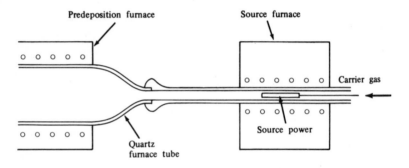

Figure 9-5 Predeposition from a powder using a source furnace.

Experiments have shown that the use of an oxide of the desired dopant produces the best results during predeposition. To guarantee that the dopant reaches the surface of the wafer as an oxide, oxygen is often introduced into the furnace tube along with the dopant.

Liquid dopant sources may be used in a set-up similar to a wet oxidation bubbler. A liquid compound containing the dopant is placed in a bubbler held at a constant temperature. A carrier gas (usually nitrogen) is bubbled through the liquid, becoming saturated with dopant, which it carries into the predeposition furnace. To obtain the oxide of the dopant on the surface of the wafer, oxygen is often introduced into the furnace tube along with the dopant-saturated carrier gas. Figure 9-6 shows a typical liquid source set-up.

Gaseous diffusion sources are also used in many applications. The use of a gaseous source simplifies the problem of introducing the dopant into the furnace, but there are several potential problems. The gases that can be easily used as dopant sources are also toxic to some

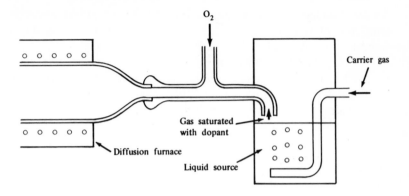

Figure 9-6   Liquid predeposition source.

degree. Care must be taken to insure that no gas leaks occur. These gases are often chemically unstable, and may decompose if improperly stored, or if stored too long. Lastly, their unstable nature limits the maximum concentration of dopant that may be present in the gas. This maximum dopant concentration means that it may be impossible to obtain a sufficiently high concentration of the wanted impurity. The dopant gases must contain the desired impurity, chemically combined with elements that do not adversely effect the predeposition process. In most cases, oxygen is added to the gas flow to guarantee that an oxide of the dopant is deposited on the wafer. A carrier gas like nitrogen may be used to keep the gas-flow velocity along the tube at the proper level. A diagram of a gaseous source system is shown in Figure 9-7.

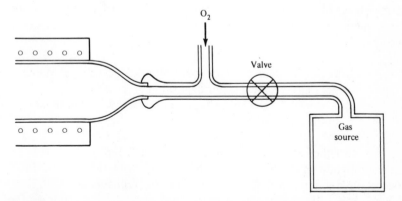

Figure 9-7   Gaseous predeposition source.

Another type of predeposition source uses wafers made of a compound of the desired dopant. The source wafers are usually prepared by oxidizing them, and both the source and the dopant wafers are placed in a quartz boat and pushed into the furnace. The boat is designed such that the silicon wafers face the source wafers and are an exact distance away from them. Usually, one source wafer has two device wafers facing it as shown in Figure 9-8.

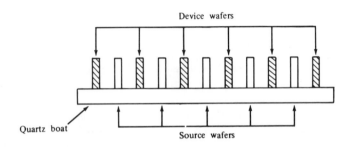

Figure 9-8 Predeposition using a source wafer.

An ambient gas, usually nitrogen, flows through the predeposition tube. A small amount of oxygen may be introduced to guarantee that the dopant that reaches the surface of the device wafers does so as an oxide. The dopant reaches the device wafers by direct transfer across the gap between the source and device wafers. A summary of powder, liquid, gaseous, and wafer sources is shown in Table 9-1.

A dopant may also be introduced into a semiconductor from a doped layer of silicon dioxide in contact with the surface of the wafer. Two methods are currently used to obtain a doped oxide layer on wafers:

1. A doped layer of oxide can be deposited using a low-temperature chemical vapor deposition process.
2. A doped layer can be "spun-on" using a technique similar to the application of photoresist.

In either case, the wafers are loaded into the furnace with the dopant already on the front side of the wafer. During the predeposition, the required amount of dopant diffuses into the semiconductor. Following this step, all surplus dopant is removed from the front side of the wafer by etching. A dilute or buffered solution of hydrofluoric acid is generally used. The wafers are now ready for the drive-in.

TABLE 9-1: Diffusion Source Chart

Type	Element	Solid Solubility (maximum)	Diffusivity at 1,100°C	Silicon Fit	Compound Name	Formula	State	Comment (NPN)
	antimony	7×10^{19} (1,250°C)	2.5×10^{-14} cm²/sec.	O.K.	antimony trioxide	Sb_2O_3	solid	subcollector
	arsenic	1.8×10^{21} (1,150°C)	3×10^{-14} cm²/sec.	good	arsenic trioxide	As_2O_3	solid	closed tube or source furnace; subcollector
n					arsine	AsH_3	gas	subcollector & emitter
	phosphorus	1.4×10^{21} (1,150°C)	3×10^{-13} cm²/sec.	average	phos pentoxide	P_2O_5	solid	emitters
					phos oxychloride	$POCl_3$	liquid	emitters
					phosphine	PH_3	gas	emitters
					silicon pyrophosphate	SiP_2O_7	solid	wafer source

TABLE 9-1 (*cont.*)

Type	Element	Solid Solubility (maximum)	Diffusivity at 1,100°C	Silicon Fit	Compound Name	Formula	State	Comment (NPN)
					boron trioxide	B_2O_3	solid	base/isolation
					boron tribromide	BBr_3	liquid	base/isolation
p	boron	5×10^{20} (1,250°C)	3×10^{-13} cm²/sec.	bad	diborane	B_2H_6	gas	base/isolation
					boron trichloride	BCl_3	gas	base/isolation
					boron nitride	BN	solid	wafer source
gold	gold	$10^{14} - 10^{17}$ (800–1,100°C)	10^{-6} cm²/sec.	good	gold	Au	solid (evap.)	base life time control
neither n nor p	iron copper lithium zinc manganese nickel							

undesirable impurities from "Contamination"

Drive-in

The drive-in is a diffusion step in which no additional dopant is introduced into the semiconductor. This process is done in an oxidizing atmosphere to regrow a protective layer of SiO_2 over the freshly diffused region. During the drive-in step, the variables of time, temperature, and ambient gases are controlled. These three variables determine:

1. The final junction depth of the diffusion.
2. The final oxide thickness over the newly doped region.
3. The exact profile of the dopant in the semiconductor.

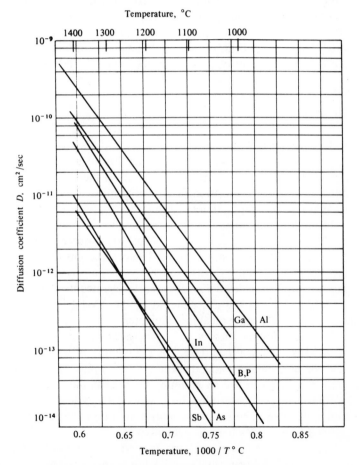

Diffusion coefficients of substitutional diffusers in silicon

Figure 9-9   Diffusion coefficients of substitutional diffusers in silicon.

A feel for the relative depth that can be obtained during the drive-in can be obtained by looking at the "diffusion coefficient" of the particular dopant in silicon. Figure 9-9 shows the diffusion coefficient of the particular dopants in silicon as a function of temperature.

9-3   DIFFUSION ANALYSIS

The two measurements most frequently used in the evaluation of a predeposition or a drive-in are its sheet resistance and its junction depth. Measurements of both of these parameters have been discussed in the section on epitaxy. The only difference between their use in epitaxial deposition and diffusion lies in the inability to determine the resistivity of a diffused layer from the sheet resistance and the junction depth. The constantly changing concentration as a function of distance from the surface makes the concept of a constant resistivity invalid in this case.

9-4   ION IMPLANTATION

A recently developed technique for introducing dopant into semiconductors is ion implantation. This process takes ions of a desired dopant, accelerates them, using an electric field, and scans them across a wafer to obtain a uniform predeposition. The energy imparted to the dopant ion determines the ion implantation depth.

The first requirement of an ion implantation system is the ability to generate ions of the desired species. A gaseous source is used with ions being generated by boiling them off. The correct ions are separated from any others by bending them through a preset angle using an electromagnetic field. The selected ions are then accelerated using an electric field and strike the target wafer penetrating the crystal lattice. Figure 9-10 shows the set-up of a typical ion implanter.

Once the doping species has been selected, the two variables that can be controlled are the "dose," or the number of ions that reach the wafer per unit of area, and the energy with which they reach the wafer. The dose is controlled by counting the ions as they pass a detector, and the energy is controlled by changing the voltage along the acceleration chamber. The ability to control both dose and energy gives rise to unique applications for this technology.

The regions implanted with the accelerated ions can be selected by using either a thick layer of silicon dioxide or, in some cases, a coating of photoresist with a pattern defined in it as a mask. The behavior of each layer as a mask is shown in Figure 9-11. Often, a thin layer of

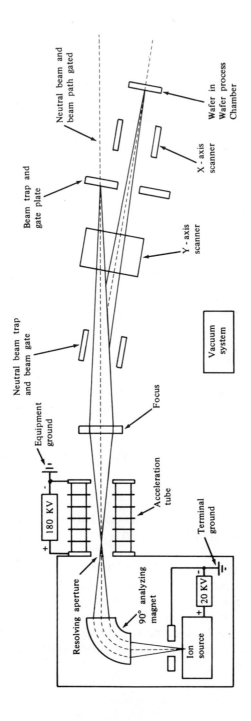

Figure 9-10 Typical ion implantation equipment.

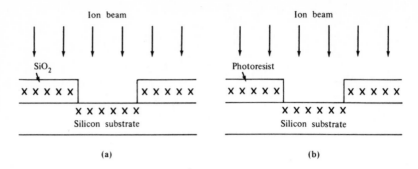

Figure 9-11   Methods of masking during ion implantation; **(a)** use of SiO$_2$ layer as a mask; **(b)** use of photoresist as a mask.

SiO$_2$ is present over the region of the wafer being implanted, and the ion penetrates this thin layer before entering the underlying substrate. Following the implantation and any subsequent cleaning steps, the implanted wafers are often put through a high-temperature furnace to activate any ions that may not have come to rest in electrically active locations in the crystal structure.

Ion implantation offers the ability to precisely control the amount of dopant introduced in regions of a semiconductor, and its depth below the surface. Control of these parameters leads to the applications for which ion implantation is most frequently used:

1. Fabrication of high-value or precise resistors
2. Control of the threshold voltage of field-effect transistors

REVIEW EXERCISES: IMPURITY INTRODUCTION AND REDISTRIBUTION I

1. Determine the solid solubility of gallium in silicon at 900°C.
2. Determine the maximum solid solubility of gold in silicon.
3. Does boron or gallium have a higher diffusion coefficient at 1,100°C?
4. Determine the diffusion coefficient of phosphorus at 1,200°C.
5. What parameter controls the penetration depth of an implanted ion?

6. During predeposition, what parameter determines concentration of dopant at the surface of a wafer?
7. What two parameters determine the predeposition profile?
8. Determine the oxide thickness necessary to selectively mask a wafer against boron diffusion at 1,100°C for 1 hour.
9. List several methods of introducing dopant impurities into a silicon wafer.
10. What three variables determine the depth of the junction during a drive-in diffusion?
11. What two measurements are frequently used to evaluate a diffusion into a silicon wafer?
12. Is it possible to obtain an accurate measure of the resistivity resulting from a diffusion? Explain.

10

Impurity Introduction and Redistribution II

10-1 MATHEMATICS OF DIFFUSION

The mathematics needed to solve for dopant profiles following either a predeposition or drive-in step appears much more difficult than it is. The derivation of the necessary formulas is beyond the scope of these lessons, but the application of the results to problems is not. The mathematics of diffusion will be studied by looking at the predeposition and the drive-in steps separately.

Predeposition

This step is performed in a high temperature diffusion furnace with an excess of the desired dopant present at the surface of the wafer. Under these conditions, the concentration of the dopant at the surface of the wafer, C_s, corresponds to the solid solubility of the dopant at the predeposition temperature. A uniform and reproducible amount of dopant enters the crystal lattice under these conditions. The profile of impurities introduced during the predeposition is found using the equation.

$$C(x) = C_s \, \text{erfc} \, \frac{x}{\sqrt{4D_1 \, t_1}} \qquad (10\text{-}1)$$

where

C_s = the solid solubility of the dopant in silicon at the predeposition temperature (from Figure 10-1).

$C(x)$ = the concentration of the dopant at a depth x into the wafer.

x = the distance from the surface of the wafer.

D_1 = the diffusion coefficient of the dopant at the predeposition temperature (from Figure 10-2)

t_1 = the time the wafers are in the predeposition

The definition of erfc follows:

$$\text{erfc}(z) = 1 - \text{erf}(z)$$

where

$$\text{erf}(z) = \frac{2}{\pi} \int_{0}^{z} e^{-a^2} d\alpha$$

Table 10-1 is a tabulation of erfc(z).

The concentration of dopant at some value of x is found in the following manner:

1. Determine the solid solubility of the dopant in the substrate material, C_s, at the predeposition temperature using Figure 10-1.
2. Determine the diffusion coefficient, D_1, of the dopant in the substrate material at the predeposition temperature using Figure 10-2.
3. Evaluate the term $\sqrt{4D_1 t_1}$ (usually the units of microns are the best ones to use).
4. Using the value of x for which the dopant concentration is being determined, evaluate the term $\dfrac{x}{\sqrt{4D_1 t_1}}$.
5. The number resulting from step 4 is Z. Using Table 10-1, the "erfc Table," determine erfc (Z).
6. The dopant concentration at a depth x is the product of C_s and erfc(Z).

This equation can be rearranged to determine the depth of a junction resulting from a predeposition as long as the resistivity or impurity concentration of the substrate is known. The junction depth, x_j, is the depth at which the dopant concentration $C(x)$ equals the background concentration C_B. Substituting x_j for x and C_B for $C(x)$ in equation 10-1 results in the equation:

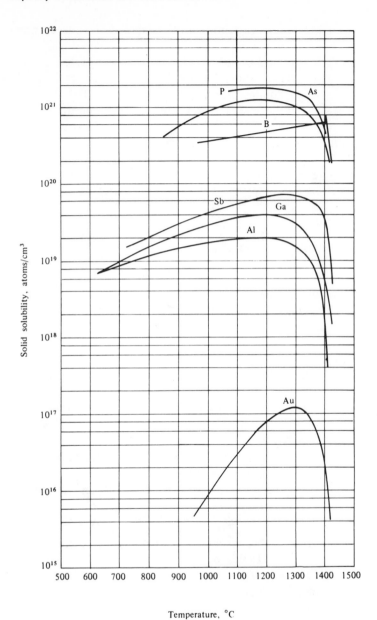

Figure 10-1 The solid solubility of impurities in silicon.

$$C_B = C_s \, \text{erfc} \; \frac{x_j}{\sqrt{4D_1 \, t_1}}$$

or

(10-2)

$$\frac{C_B}{C_s} = \text{erfc} \; \frac{x_j}{\sqrt{4D_1 \, t_1}}$$

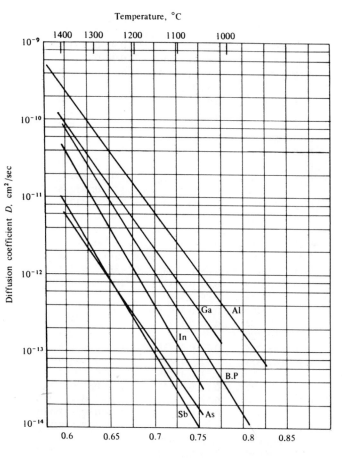

Figure 10-2 Diffusion coefficients of substitutional diffusers.

TABLE 10-1: Complementary Error Function

Z	erfc(Z)	Z	erfc(Z)	Z	erfc(Z)
0.00	1.00000	0.43	0.543113	0.86	0.223900
0.01	0.988717	0.44	0.533775	0.87	0.218560
0.02	0.977435	0.45	0.524518	0.88	0.213313
0.03	0.966159	0.46	0.515345	0.89	0.208157
0.04	0.954889	0.47	0.506255	0.90	0.203092
0.05	0.943628	0.48	0.497250		
0.06	0.932378	0.49	0.488332	0.91	0.198117
0.07	0.921142	0.50	0.479500	0.92	0.193232
0.08	0.909922			0.93	0.188436
0.09	0.898719	0.51	0.470756	0.94	0.183729
0.10	0.887537	0.52	0.462101	0.95	0.179109
		0.53	0.453536	0.96	0.174576
0.11	0.876377	0.54	0.445061	0.97	0.170130
0.12	0.865242	0.55	0.436677	0.98	0.165768
0.13	0.854133	0.56	0.428384	0.99	0.161492
0.14	0.843053	0.57	0.420184	1.00	0.157299
0.15	0.832004	0.58	0.412077		
0.16	0.820988	0.59	0.404063	1.01	0.153190
0.17	0.810008	0.60	0.396144	1.02	0.149162
0.18	0.799064			1.03	0.145216
0.19	0.788160	0.61	0.388319	1.04	0.141350
0.20	0.777297	0.62	0.380589	1.05	0.137564
		0.63	0.372954	1.06	0.133856
0.21	0.766478	0.64	0.365414	1.07	0.130227
0.22	0.755704	0.65	0.357971	1.08	0.126674
0.23	0.744977	0.66	0.350623	1.09	0.123197
0.24	0.734300	0.67	0.343372	1.10	0.119795
0.25	0.723674	0.68	0.336218		
0.26	0.713100	0.69	0.329160	1.11	0.116467
0.27	0.702582	0.70	0.322199	1.12	0.113212
0.28	0.692120			1.13	0.110029
0.29	0.681716	0.71	0.315334	1.14	0.106918
0.30	0.671373	0.72	0.308567	1.15	0.103876
		0.73	0.301896	1.16	0.100904
0.31	0.661092	0.74	0.295322	1.17	0.979996D-01
0.32	0.650874	0.75	0.288844	1.18	0.951626D-01
0.33	0.640721	0.76	0.282463	1.19	0.923917D-01
0.34	0.630635	0.77	0.276178	1.20	0.896860D-01
0.35	0.620618	0.78	0.269990		
0.36	0.610670	0.79	0.263897	1.21	0.870445D-01
0.37	0.600794	0.80	0.257899	1.22	0.844661D-01
0.38	0.590990			1.23	0.819499D-01
0.39	0.581261	0.81	0.251997	1.24	0.794948D-01
0.40	0.571608	0.82	0.246189	1.25	0.770999D-01
		0.83	0.240476	1.26	0.747640D-01
0.41	0.562031	0.84	0.234857	1.27	0.724864D-01
0.42	0.552532	0.85	0.229332	1.28	0.702658D-01

TABLE 10-1 (*cont.*)

Z	erfc(Z)	Z	erfc(Z)	Z	erfc(Z)
1.29	0.681014D-01	1.72	0.149972D-01	2.15	0.236139D-02
1.30	0.659920D-01	1.73	0.144215D-01	2.16	0.225285D-02
1.31	0.639369D-01	1.74	0.138654D-01	2.17	0.214889D-02
1.32	0.619348D-01	1.75	0.133283D-01	2.18	0.204935D-02
1.33	0.599850D-01	1.76	0.128097D-01	2.19	0.195406D-02
1.34	0.580863D-01	1.77	0.123091D-01	2.20	0.186285D-02
1.35	0.562378D-01	1.78	0.118258D-01		
1.36	0.544386D-01	1.79	0.113594D-01	2.21	0.177556D-02
1.37	0.526876D-01	1.80	0.109095D-01	2.22	0.169205D-02
1.38	0.509840D-01			2.23	0.161217D-02
1.39	0.493267D-01	1.81	0.104755D-01	2.24	0.153577D-02
1.40	0.477149D-01	1.82	0.100568D-01	2.25	0.146272D-02
1.41	0.461476D-01	1.83	0.965319D-02	2.26	0.139288D-02
1.42	0.446238D-01	1.84	0.926405D-02	2.27	0.132613D-02
1.43	0.431427D-01	1.85	0.888897D-02	2.28	0.126234D-02
1.44	0.417034D-01	1.86	0.852751D-02	2.29	0.120139D-02
1.45	0.403050D-01	1.87	0.817925D-02	2.30	0.114318D-02
1.46	0.389465D-01	1.88	0.784378D-02		
1.47	0.376271D-01	1.89	0.752068D-02	2.31	0.108758D-02
1.48	0.363459D-01	1.90	0.720957D-02	2.32	0.103449D-02
1.49	0.351021D-01			2.33	0.983805D-03
1.50	0.338949D-01	1.91	0.691006D-02	2.34	0.935430D-03
1.51	0.327233D-01	1.92	0.662177D-02	2.35	0.889267D-03
1.52	0.315865D-01	1.93	0.634435D-02	2.36	0.845223D-03
1.53	0.304838D-01	1.94	0.607743D-02	2.37	0.803210D-03
1.54	0.294143D-01	1.95	0.582066D-02	2.38	0.763142D-03
1.55	0.283773D-01	1.96	0.557372D-02	2.39	0.724936D-03
1.56	0.273719D-01	1.97	0.533627D-02	2.40	0.688514D-03
1.57	0.263974D-01	1.98	0.510800D-02		
1.58	0.254530D-01	1.99	0.488859D-02	2.41	0.653798D-03
1.59	0.245380D-01	2.00	0.467773D-02	2.42	0.620716D-03
1.60	0.236516D-01			2.43	0.589197D-03
1.61	0.227932D-01	2.01	0.447515D-02	2.44	0.559174D-03
1.62	0.219619D-01	2.02	0.428055D-02	2.45	0.530580D-03
1.63	0.211572D-01	2.03	0.409365D-02	2.46	0.503353D-03
1.64	0.203782D-01	2.04	0.391419D-02	2.47	0.477434D-03
1.65	0.196244D-01	2.05	0.374190D-02	2.48	0.452764D-03
1.66	0.188951D-01	2.06	0.357654D-02	2.49	0.429288D-03
1.67	0.181896D-01	2.07	0.341785D-02	2.50	0.406952D-03
1.68	0.175072D-01	2.08	0.326559D-02		
1.69	0.168474D-01	2.09	0.311954D-02	2.51	0.385705D-03
1.70	0.162095D-01	2.10	0.297947D-02	2.52	0.365499D-03
				2.53	0.346286D-03
1.71	0.155930D-01	2.11	0.284515D-02	2.54	0.328021D-03
		2.12	0.271639D-02	2.55	0.310660D-03
		2.13	0.259298D-02	2.56	0.294163D-03
		2.14	0.247471D-02	2.57	0.278489D-03

TABLE 10-1 (*cont.*)

Z	erfc(Z)	Z	erfc(Z)	Z	erfc(Z)
2.58	0.263600D-03	3.01	0.207390D-04	3.44	0.114518D-05
2.59	0.249461D-03	3.02	0.194664D0-4	3.45	0.106605D-05
2.60	0.236034D-03	3.03	0.182684D-04	3.46	0.992201D-06
		3.04	0.171409D-04	3.47	0.923288D-06
2.61	0.223289D-03	3.05	0.160798D-04	3.48	0.858995D-06
2.62	0.211191D-03	3.06	0.150816D-04	3.49	0.799025D-06
2.63	0.199711D-03	3.07	0.141426D-04	3.50	0.743098D-06
2.64	0.188819D103	3.08	0.132595D-04		
2.65	0.178488D-03	3.09	0.124292D-04	3.51	0.690952D-06
2.66	0.168689D-03	3.10	0.116487D-04	3.52	0.642341D-06
2.67	0.159399D-03			3.53	0.597035D-06
2.68	0.150591D-03	3.11	0.109150D-04	3.54	0.554816D-06
2.69	0.142243D-03	3-12	0.102256D-04	3.55	0.515484D-06
2.70	0.134333D-03	3-13	0.957795D-05	3.56	0.478847D-06
		3.14	0.896956D-05	3.57	0.444728D-06
2.71	0.126838D-03	3.15	0.839821D-05	3.58	0.412960D-06
2.72	0.119738D-03	3.16	0.786174D-05	3.59	0.383387D-06
2.73	0.113015D-03	3.17	0.735813D-05	3.60	0.355863D-06
2.74	0.106649D-03	3.18	0.688545D-05		
2.75	0.100622D-03	3.19	0.644190D-05	3.61	0.330251D-06
2.76	0.949176D-04	3.20	0.602576D-05	3.62	0.306423D-06
2.77	0.895197D-04			3.63	0.284259D-06
2.78	0.844127D-04	3.21	0.563542D-05	3.64	0.263647D-06
2.79	0.795818D-04	3.22	0.526935D-05	3.65	0.244483D-06
2.80	0.750132D-04	3.23	0.492612D-05	3.66	0.226667D-06
		3.24	0.460435D-05	3.67	0.210109D-06
2.81	0.706933D-04	3.25	0.430278D-05	3.68	0.194723D-06
2.82	0.666096D-04	3.26	0.402018D-05	3.69	0.180429D-06
2.83	0.627497D-04	3.27	0.375542D-05	3.70	0.167151D-06
2.84	0.591023D-04	3.28	0.350742D-05		
2.85	0.556563D-04	3.29	0.327517D-05	3.71	0.154821D-06
2.86	0.524012D-04	3.30	0.305771D-05	3.72	0.143372D-06
2.87	0.493270D-04			3.73	0.132744D-06
2.88	0.464244D-04	3.31	0.285414D-05	3.74	0.122880D-06
2.89	0.436842D-04	3.32	0.266360D-05	3.75	0.113727D-06
2.90	0.410979D-04	3.33	0.248531D-05	3.76	0.105236D-06
		3.34	0.231850D-05	3.77	0.973591D-07
2.91	0.386573D-04	3.35	0.216248D-05	3.78	0.900547D-07
2.92	0.363547D-04	3.36	0.201656D-05.	3.79	0.832821D-07
2.93	0.341828D-04	3.37	0.188013D-05	3.80	0.770039D-07
2.94	0.321344D-04	3.38	0.175259D-05		
2.95	0.302030D-04	3.39	0.163338D-05	3.81	0.711851D-07
2.96	0.283823D-04	3.40	0.152199D-05	3.82	0.657933D-07
2.97	0.266662D-04			3.83	0.607981D-07
2.98	0.250491D-04	3.41	0.141793D-05	3.84	0.561711D-07
2.99	0.235256D-04	3.42	0.132072D-05	3.85	0.518863D-07
3.00	0.220905D-04	3.43	0.122994D-05	3.86	0.479189D-07

TABLE 10-1 (*cont.*)

Z	erfc(Z)	Z	erfc(Z)	Z	erfc(Z)
3.87	0.442464D-07	4.30	0.119347D-08	4.73	0.224348D-10
3.88	0.408473D-07			4.74	0.203664D-10
3.89	0.377021D-07	4.31	0.109259D-08	4.75	0.184850D-10
3.90	0.347922D-07	4.32	0.100005D-08	4.76	0.167742D-10
		4.33	0.915161D-09	4.77	0.152187D-10
3.91	0.321007D-07	4.34	0.837317D-09	4.78	0.138048D-10
3.92	0.296117D-07	4.35	0.765944D-09	4.79	0.125198D-10
3.93	0.273103D-07	4.36	0.700518D-09	4.80	0.113521D-10
3.94	0.251829D-07	4.37	0.640556D-09		
3.95	0.232167D-07	4.38	0.585612D-09	4.81	0.102914D-10
3.96	0.213999D-07	4.39	0.535276D-09	4.82	0.932791D-11
3.97	0.197214D-07	4.40	0.489171D-09	4.83	0.845298D-11
3.98	0.181710D-07			4.84	0.765861D-11
3.99	0.167392D-07	4.41	0.446950D-09	4.85	0.693754D-11
4.00	0.154173D-07	4.42	0.408293D-09	4.86	0.628312D-11
		4.43	0.372906D-09	4.87	0.568932D-11
4.01	0.141969D-07	4.44	0.340520D-09	4.88	0.515062D-11
4.02	0.130707D-07	4.45	0.310886D-09	4.89	0.466202D-11
4.03	0.120314D-07	4.46	0.283775D-09	4.90	0.421893D-11
4.04	0.110726D-07	4.47	0.258978D-09		
4.05	0.101882D-07	4.48	0.236302D-09	4.91	0.381721D-11
4.06	0.937269D-08	4.49	0.215568D-09	4.92	0.345307D-11
4.07	0.862073D-08	4.50	0.196616D-09	4.93	0.312304D-11
4.08	0.792756D-08			4.94	0.282401D-11
4.09	0.728870D-08	4.51	0.179295D-09	4.95	0.255311D-11
4.10	0.670003D-08	4.52	0.163467D-09	4.96	0.230774D-11
		4.53	0.149008D-09	4.97	0.208554D-11
4.11	0.615769D-08	4.54	0.135801D-09	4.98	0.188437D-11
4.12	0.565816D-08	4.55	0.123740D-09	4.99	0.170226D-11
4.13	0.519813D-08	4.56	0.112729D-09	5.00	0.153746D-11
4.14	0.477457D-08	4.57	0.102677D-09		
4.15	0.438468D-08	4.58	0.935034D-10	5.01	0.138834D-11
4.16	0.402583D-08	4.59	0.851326D-10	5.02	0.125343D-11
4.17	0.369564D-08	4.60	0.774960D-10	5.03	0.113141D-11
4.18	0.339186D-08			5.04	0.102107D-11
4.19	0.311245D-08	4.61	0.705306D-10	5.05	0.921310D-12
4.20	0.285549D-08	4.62	0.641787D-10	5.06	0.831132D-12
		4.63	0.583874D-10	5.07	0.749634D-12
4.21	0.261924D-08	4.64	0.531083D-10	5.08	0.675994D-12
4.22	0.240207D-08	4.65	0.482970D-10	5.09	0.609469D-12
4.23	0.220247D-08	4.66	0.439130D-10	5.10	0.549382D-12
4.24	0.201907D-08	4.67	0.399191D-10		
4.25	0.185057D-08	4.68	0.362814D-10	5.11	0.495122D-12
4.26	0.169581D-08	4.69	0.329687D-10	5.12	0.446133D-12
4.27	0.155369D-08	4.70	0.299526D-10	5.13	0.401912D-12
4.28	0.142319D-08	4.71	0.272071D-10	5.14	0.362004D-12
4.29	0.130341D-08	4.72	0.247084D-10	5.15	0.325994D-12

TABLE 10-1 (*cont.*)

Z	erfc(Z)	Z	erfc(Z)	Z	erfc(Z)
5.16	0.293508D-12	5.44	0.143363D-13	5.72	0.600078D-15
5.17	0.264208D-12	5.45	0.128342D-13	5.73	0.534249D-15
5.18	0.237786D-12	5.46	0.114873D-13	5.74	0.475548D-15
5.19	0.213964D-12	5.47	0.102797D-13	5.75	0.423213D-15
5.20	0.192491D-12	5.48	0.919719D-14	5.76	0.376564D-15
		5.49	0.822708D-14	5.77	0.334990D-15
5.21	0.173138D-12	5.50	0.735785D-14	5.78	0.297948D-15
5.22	0.155701D-12			5.79	0.264949D-15
5.23	0.139992D-12	5.51	0.657916D-14	5.80	0.235559D-15
5.24	0.125844D-12	5.52	0.588172D-14		
5.25	0.113103D-12	5.53	0.525717D-14	5.81	0.209387D-15
5.26	0.101632D-12	5.54	0.469802D-14	5.82	0.186087D-15
5.27	0.913067D-13	5.55	0.419751D-14	5.83	0.165347D-15
5.28	0.820141D-13	5.56	0.374959D-14	5.84	0.146889D-15
5.29	0.736527D-13	5.57	0.334880D-14	5.85	0.130466D-15
5.30	0.661308D-13	5.58	0.299027D-14	5.86	0.115856D-15
		5.59	0.266959D-14	5.87	0.102862D-15
5.31	0.593654D-13	5.60	0.238284D-14	5.88	0.913078D-16
5.32	0.532816D-13			5.89	0.810352D-16
5.33	0.478119D-13	5.61	0.212646D-14	5.90	0.719040D-16
5.34	0.428952D-13	5.62	0.189730D-14		
5.35	0.384766D-13	5.63	0.169250D-14	5.91	0.637892D-16
5.36	0.345063D-13	5.64	0.150951D-14	5.92	0.565791D-16
5.37	0.309396D-13	5.65	0.134604D-14	5.93	0.501740D-16
5.38	0.277362D-13	5.66	0.120003D-14	5.94	0.444852D-16
5.39	0.248595D-13	5.67	0.106965D-14	5.95	0.394336D-16
5.40	0.222768D-13	5.68	0.953249D-15	5.96	0.349488D-16
		5.69	0.849347D-15	5.97	0.309679D-16
5.41	0.199585D-13	5.70	0.756621D-15	5.98	0.274350D-16
5.42	0.178779D-13			5.99	0.243004D-16
5.43	0.160110D-13	5.71	0.673885D-15		

The junction depth x_j can be determined by following these steps:

1. Evaluate $\dfrac{C_B}{C_s}$. This value is erfc(Z).

2. Using the erfc Table, determine the value of Z that results in erfc(Z) $= \dfrac{C_B}{C_s}$.

3. This value of Z must equal $\dfrac{x_j}{\sqrt{4D_1 t_1}}$. Setting them equal, $Z = \dfrac{x_j}{\sqrt{4D_1 t_1}}$ or $x_j = Z\sqrt{4D_1 t_1}$.

4. Determine x_j using D_1, the diffusion coefficient of the impurity at the predeposition temperature, and t_1, the predeposition time.

The total amount of impurity, Q, introduced during a predeposition is found by evaluating the expression:

$$Q = C_s\sqrt{\frac{4D_1t_1}{\pi}} \text{ (atoms/cm}^2) \tag{10-3}$$

where C_s, D_1, and t_1 correspond to the values defined above.

Drive-in

The predeposition has introduced a precise amount of impurity into the crystal lattice, but the junction depth and resulting profile are often not adequate to produce the desired semiconductor devices. The final junction depth and impurity profile are produced by a drive-in operation. A drive-in is performed in a high temperature diffusion furnace once the excess dopant remaining on the surface of the wafer from the predeposition has been removed by an etch step.

If the impurity profile resulting from the predeposition is approximated by a rectangle that is very tall and narrow (often called a delta function) the drive-in results in a dopant profile expressed by the equation:

$$C(x) = \left(\frac{Q}{\sqrt{\pi D_2 t_2}}\right)e^{-(x^2)/(4D_2t_2)} \tag{10-4}$$

where $C(x)$ = the dopant concentration a distance x from the surface of the wafer.

Q = the amount of dopant introduced into the crystal during the predeposition (Equation 10-3).

D_2 = The diffusion of the dopant at the drive-in temperature.

t_2 = the drive-in time.

e = constant = 2.71828

x = the depth into the wafer.

This approximation is valid for most predepositions

Equation 10-3 can be used to determine the dopant concentration a distance x from the surface as follows:

1. Determine Q from the predeposition conditions.
2. Determine D_2 and t_2 from the drive-in conditions.
3. Evaluate $\dfrac{Q}{\sqrt{\pi D_2 t_2}}$ and $\dfrac{1}{(4D_2 t_2)}$
4. For the chosen x, evaluate: $e^{-(x^2)/(4D_2 t_2)}$
5. Evaluate $C(x)$.

This equation can also be used to determine the junction depth resulting from a drive-in if the background concentration of impurities in the substrate C_B is known. As in the predeposition step, the junction depth x_j is the depth x at which the background concentration equals the dopant concentration. Substituting these values in Equation 10-4 we have:

$$C_B = \left(\frac{Q}{\sqrt{\pi D_2 t_2}}\right) e^{-(x_j^2)/(4D_2 t_2)}$$

or

$$\frac{C_B \sqrt{\pi D_2 t_2}}{Q} = e^{-(x_j^2)/(4D_2 t_2)} \qquad (10\text{-}5)$$

The value of x_j can be determined by following these steps:

1. Evaluate the expression on the left of the equal sign.
$$\frac{C_B \sqrt{\pi D_2 t_2}}{Q}$$

2. Determine the exponent, W, that solves the equation
$$e^{-W} = \frac{C_B \sqrt{\pi D_2 t_2}}{Q}$$

3. Set W equal to $\dfrac{x_j}{(4D_2 t_2)}$ and solve for x_j. $x_j = \sqrt{4D_2 t_2 W}$

Examples The mathematics just presented may seem like a lot of work, but it is not too difficult if done carefully. To demonstrate the use of this information, let's consider a predeposition and then a drive-in example:

A. Predeposition: A predeposition is performed on a silicon wafer at 975°C for 30 minutes in the presence of an excess of phosphorus.

1. The concentration of phosphorus as a function of depth can be determined as shown below:

   a. From Figure 10-1, $C_s = 8 \times 10^{20}$ atoms/cm^3

   b. From Figure 10-2, $D_1 = 1.7 \times 10^{-14}$ cm^2/sec

   c. $t_1 = 30$ minutes $= 1800$ seconds

   $$\sqrt{4D_1t_1} = \sqrt{(4)\,(1.7 \times 10^{-14})\,(1800)}$$
   $$= \sqrt{1.22 \times 10^{-10}\text{ cm}^2} = 1.106 \times 10^{-5}\text{cm}$$

   But, 1.106×10^{-5} cm $= .1106\mu \cong .11\mu$

   Therefore $C(x) = C_s$ erfc $\left(\dfrac{x}{\sqrt{4D_1t_1}}\right)$

   $$= C_s \text{ erfc } \left(\frac{x}{.11\mu}\right) = C_s \text{ erfc}(z)$$

   Solve for z and erfc(z), and $C(x)$ as shown in Table 10-2. The dopant distribution resulting from this predeposition is shown in Figure 10-3.

TABLE 10-2: Solution to the Predeposition Problem

x	z	erfc(Z)	$C(x)$
0	0	1	8 $\times 10^{20}$/cm^3
.1μ	.9042	.20	1.6 $\times 10^{20}$/cm^3
.2μ	1.8083	.010	8 $\times 10^{18}$/cm^3
.3μ	2.7125	1.23 $\times 10^{-4}$	9.8 $\times 10^{16}$/cm^3
.4μ	3.6166	3.1 $\times 10^{-7}$	2.48 $\times 10^{14}$/cm^3
.5μ	4.5208	1.6 $\times 10^{-10}$	1.28 $\times 10^{11}$/cm^3

2. The junction depth is determined by the point of transition from n-type to p-type silicon. If this predeposition was done using

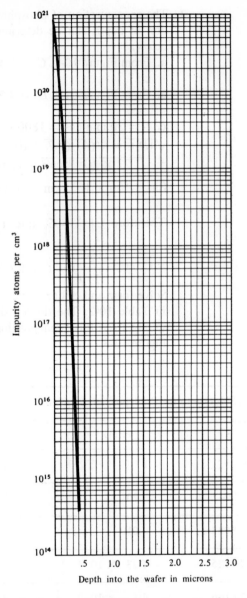

Figure 10-3 The distribution of phosphorus atoms in silicon following the predeposition step.

a .3 Ω-cm p-type wafer, the junction depth can be determined from either Figure 10-3, or by solving the proper equation. A .3 Ω-cm p-type substrate corresponds to a background concentration C_B of 10^{17} atoms/cm^3. From Figure 10-3, a C_B of 10^{17} atoms/cm^3 would result in a junction depth x_j of about .3μ. Mathematically,

$$C_B = C_s \, \mathrm{erfc} \, (\frac{x_j}{\sqrt{4D_1 t_1}})$$

where $\quad C_B = 10^{17}$ atoms/cm^3

$$\mathrm{erfc} \, (\frac{x_j}{.1106}) = C_B/C_s = 10^{17}/8 \times 10^{20}$$
$$= 1.25 \times 10^{-4}$$

Therefore, $\dfrac{x_j}{.1106} = 2.71$ from Table 10-1,

and $x_j = (.1106) \, (2.71) \cong .3\mu$.

3. The total amount of dopant incorporated into the wafer per unit area can be determined from the equation

$$Q = C_s \, \sqrt{\frac{4D_1 t_1}{\pi}}$$

$$= (8 \times 10^{20}) \, \sqrt{\frac{.1106}{\pi}}$$

atoms/cm^2
$Q = 5 \times 10^{15}$ atoms/cm^2

B. Drive-in: Using the wafer from A above, determine the concentration as a function of depth following a 50-minute drive-in at 1,100°C.

1. The dopant profile is determined as follows:
   a. From Figure 10-2, $D_2 = 3.3 \times 10^{-13}$ cm^2/sec
   b. $t_2 = 50$ minutes $= 3,000$ seconds

$$C(x) = \left(\frac{Q}{\sqrt{\pi D_2 t_2}}\right) e^{-(x^2)/(4D_2 t_2)}$$

$$= \left(\frac{5 \times 10^{15}/\mathrm{cm}^2}{5.58 \times 10^{-5}\mathrm{cm}}\right) e^{-(x^2)/(4D_2 t_2)}$$

$$C(x) = \left(\frac{9 \times 10^{19}}{\text{cm}^3}\right) e^{-(x^2)/(4D_2 t_2)}$$

This equation is solved for values of x in Table 10-3. The dopant profile resulting from this drive-in is shown in Figure 10-4.

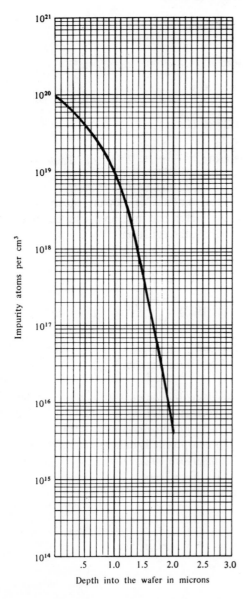

Impurity atoms per cm³

Depth into the wafer in microns

Figure 10-4 The distribution of phosphorus atoms in silicon following the drive-in step.

2. The junction depth of the dopant following the drive-in depends on the resistivity of the initial wafer, but for .3 Ω-cm p-type material as used in the predeposition, Figure 10-4 shows a junction depth of about 1.65 microns.

TABLE 10-3: Solution to the Drive-in Problem

$x(\mu)$	$x^2(\mu)$	$x^2/4D_2t_2$	$e^{-(x^2)/(4D_2t_2)}$	$C(x)(/cm^3)$
0	0	0	1	9×10^{19}
.5	.25	.63	.53	4.77×10^{19}
1.0	1.0	2.5	.0821	7.34×10^{18}
1.5	2.25	5.63	3.54×10^{-3}	3.23×10^{17}
2.0	4.0	10	4.54×10^{-5}	4.05×10^{15}
2.5	6.25	15.6	1.6×10^{-7}	1.48×10^{13}
3.0	9.0	22.5	1.69×10^{-10}	1.52×10^{10}

REVIEW EXERCISES: IMPURITY INTRODUCTION AND REDISTRIBUTION II

1. Using Table 10-1, determine:
   a. The erfc of 4.53
   b. The number whose erfc is 3.57654×10^{-3}
2. Following the predeposition in the example, at what depth would a junction be present if the step was done using a p-type substrate doped with:
   a. 5×10^{16} atoms/cm^3
   b. 5×10^{19} atoms/cm^3

3. Following the drive-in step in the example, determine the junction depth if the step was done using a p-type substrate doped with:
   a. 5×10^{16} atoms/cm^3
   b. 5×10^{19} atoms/cm^3

4. What are the impurity profiles after the (a) predeposition (b) drive-in diffusion?

5. For a normal diffusion process does the surface resistivity (as measured with a four-point probe) depend directly or inversely on the initial amount of the predeposition (Q)? Explain.

6. As time progresses during a predeposition will the surface resistivity increase or decrease? Explain.

11

Photomasking

11-1 INTRODUCTION

Photomasking can be divided into two distinct areas, both of which are necessary for the successful transfer of an image to the surface of a semiconductor wafer. These two areas are:

1. The generation of the "mask" whose image is transferred to the wafer.
2. The process of transferring the image from the mask to the surface of a wafer through the use of a sensitized layer often called photoresist.

Both of these processes will be discussed to give a better idea of the entire area of photomasking.

11-2 GENERATION OF A PHOTOMASK

The first step in the long journey to a completed integrated circuit is the successful completion of a test circuit or a test "breadboard." A breadboard is put together and tested by an engineer to determine whether his ideas for a new circuit work. The breadboard is constructed from discrete components or from parts of other circuits connected together. Extensive tests are run on the breadboard to determine its operating characteristics as temperature, supply voltage, and other parameters vary. Once the breadboard has been thoroughly tested, the translation to the many layers of a photomask begins.

The circuit will eventually be fabricated by sequentially transferring images to the front side of a wafer while performing steps such as chemical vapor deposition, epitaxy, predeposition, and drive-in, or metallization between successive image transfers. Figure 11-1 shows the masks that are transferred to a wafer during a seven-mask process.

Mask layout is the task of converting the circuit schematic to the final device layout. It consists of the following steps:

1. Drawing geometrics representing all of the devices in the circuit.
2. Arranging these components to occupy a minimum of space while making the device interconnection and connection to the outside world as easy as possible.
3. Breaking this composite drawing down into layers for subsequent processing.

These three steps may be automated to varying degrees through the use of computer-controlled drawing boards and other equipment, but these aids simply make it easier to design while leaving the creative task of the layout to the operator. Following these three steps, the actual task of mask making begins.

Copies of the layers produced above are made using either manual or computer techniques and are photographically reduced until they are ten times the ultimate size. Use of a step-and-repeat camera and the 10x plate results in rows and columns of the identical image being transferred to a glass plate called a "master." A master plate of each layer is produced in this manner. Next the master is used to manufacture a "submaster," again using a photosensitized glass plate. Finally, many copies or working plates of each submaster are made using more photosensitized glass plates. It is these working plates that are used in the actual transfer of the image to the front side of a semiconductor wafer.

Though glass plates covered with photosensitive emulsion are often used in all steps of mask preparation, the emulsion is susceptible to scratches and tears. Alternate materials sometimes substituted for emulsion include chrome, silicon, and iron oxide. All three of these materials stand up to wear better than emulsion masks, but are considerably more expensive. Iron oxide and silicon masks have the additional advantage of being transparent to the yellow light used to align the masks, while being opaque to the intense ultraviolet light used for the exposure. Visually, each type of mask is a plate of glass with alternate clear and opaque regions as shown in Figure 11-2.

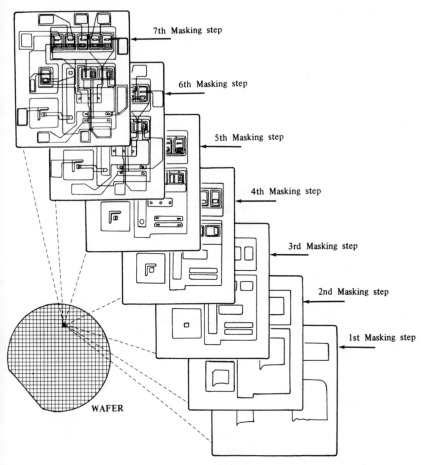

7th Masking step

6th Masking step

5th Masking step

4th Masking step

3rd Masking step

2nd Masking step

1st Masking step

WAFER

Figure 11-1   The layers transferred to a wafer during a seven-mask process.

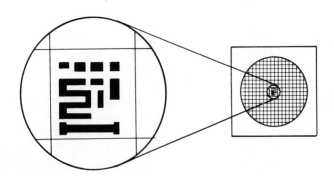

Figure 11-2   A working plate mask.

11-3 PHOTOLITHOGRAPHY

The photolithographic process is the transfer of an image from the mask to a wafer through the use of a photosensitive material often called photoresist. Photoresist is a chemical formulation containing a light-sensitive material in suspension in a solvent. The light-sensitive material has been selected so that it responds to the intense blue-violet light put out by a mercury arc lamp, but does not respond to the red or yellow light commonly in use in darkrooms or photoresist areas. Photoresist comes in two distinct types:

1. *Light-hardened resist.* The light from the exposure process hardens or polymerizes the photoresist. They are usually called negative resists.
2. *Light-softened resist.* The light from the exposure process softens or depolymerizes the photoresist. They are usually called positive resists.

Either type of resist may be used for all applications, but in certain situations it is advantageous to use one over the other.

A photoresist is characterized by four parameters that affect its performance. These are:

1. *Adhesion.* A measure of the lateral etch at the edge of a post-baked resist image.
2. *Etch resistance.* An oxidized wafer fully coated with photoresist is subjected to an etch several times longer than normal, and any breakdown in the photoresist is looked for.
3. *Resolution.* The minimum bar width whose image can be successfully transferred to the resist layer is measured.
4. *Photosensitivity.* The absolute response to different light intensities is measured.

These tests are often performed by a manufacturer to maintain a constant production quality. These tests should be performed by the user on all lots of photoresist to verify that the high quality photoresist necessary for today's manufacturing is always being used.

The amount of solvent in a batch of photoresist determines its thickness or viscosity. The more viscous a resist, the less easily it flows. Honey is a more viscous material than water because it does not spread out as fast as water from a drop on a surface. The viscosity of a photoresist is measured in units of either centipoise or centistoke. These units are closely related, but are not the same. Most photoresists are used in the 28–60 centipoise range which means that they flow about like syrup.

The photolithographic process consists of a number of steps performed sequentially, regardless of the particular photoresist being used or the layer to which it is being applied. These steps are shown in the photoresist flowchart.

BASIC PHOTORESIST FLOWCHART

STEP   OPERATION

1. SUBSTRATE PREPARATION: oxidation, CVD, etc.
2. SURFACE PREPARATION: clean, dehydrate, prime, etc.
3. APPLICATION OF RESIST: spin, spray, roll, dip, etc.
4. SOFT BAKE: low temperature cure to dry resist.
5. EXPOSE: align and expose to selectively polymerize the resist.
6. DEVELOP: dissolve the unpolymerized resist.
7. VISUAL INSPECTION (develop check): verify accurate image transfer to the photoresist.
8. HARD BAKE: higher temperature cure to completely dry and polymerize the resist.
9. ETCH: oxide, metal, etc.
10. STRIP RESIST: organic, asher, or acid removal of resist.
11. VISUAL INSPECTION (final inspection): verify accurate image transfer to the layer.

The surface preparation required by a substrate depends on the last operation the substrate has seen. In many cases, such as with wafers just removed from a diffusion or oxidation furnace or from a metal evaporator, no preparation is needed. Some surfaces such as silicon nitride or polycrystalline silicon may require surface preparation. Oxidation of both of these materials is a common technique, and is performed as described in the section on oxidation. Another technique that is sometimes used is called priming. The use of a priming solution increases the adhesion of the photoresist to the surface. Primer may be applied by immersing the substrates in the priming solution, spraying the solution on, or by passing gas laden with priming vapors over the surface of the wafer. Application of some primers requires the baking of the substrates before subsequent coating with photoresist.

Photoresist may be applied using a variety of techniques including dipping, spraying, brushing, or rollercoating, but in the fabrication of semiconductor devices, the most satisfactory technique is the use of a "spinner." A cross section of a spinner with a wafer on it is shown in Figure 11-3.

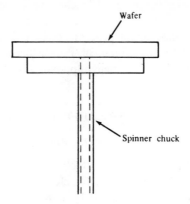

Figure 11-3 Cross section of a spinner with a wafer on it.

It consists of a wafer holder on a hollow, rotating shaft. A vacuum is used to hold the wafer on the holder while it is in motion. A precise amount of photoresist is dispensed onto the center of the wafer and the wafer begins to spin. The photoresist on the wafer moves outward, uniformly coating the wafer. Any excess photoresist is spun off of the edge of the wafer. The spin speed and the viscosity of the resist determine the thickness of the photoresist following application. Figure 11-4 shows the thickness of photoresist as a function of spin speed for different viscosity resists.

A photoresist will have specified minimum and a maximum spin speeds for obtaining uniform layers. If too low a spin speed is used, an excessive edge bead forms. The use of too high a spin speed produces a nonuniform layer because of uneven evaporation of the solvent in the resist.

Following the resist application, excess solvents are baked out of the resist during the soft-bake step. Two methods of baking the resist are in common use:

1. *Forced Hot Air.* A circulating current of hot air removes excess solvent from the resist.
2. *Infrared (IR).* The heat produced by special infrared light bulbs heats the wafers, thus evaporating the excess solvent.

Temperature and time are the two major control variables. Baking at too low a temperature requires excessive time, while baking at too high a temperature results in the surface being sealed while solvent is still present in lower levels. This condition leads to a wrinkled appearance in the resist surface.

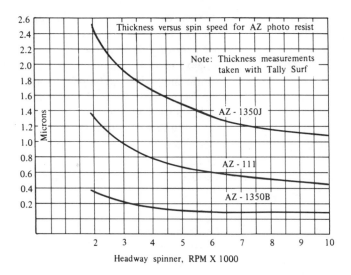

Figure 11-4   Photoresist thickness vs. spin speed.

After wafers have been cooled for a sufficient amount of time following the soft bake, they are ready for the alignment step. Using a precise piece of optical/mechanical equipment, the mask and wafer are brought close together, and the image on the mask is aligned to any pattern already existing in the wafer beneath the layer of photoresist. For the first mask, no alignment is necessary. Alignment for successive layers is made possible by the use of a microscope and precision controls for positioning the wafer with respect to the mask. Once the alignment is done, a high intensity ultraviolet (UV) mercury arc light source shines through the mask, exposing the resist in places not protected by opaque regions of the mask.

Immediately following the exposure step, the unpolymerized regions of the resist are dissolved at the develop step. The develop step may work by immersing the wafer in the developer, spraying it on, or atomizing it. Atomizing the developer uses a minimum amount of developer, and is favored in many applications. The develop step should leave a sharp edge where the photoresist stops. A rinse is usually applied following the developer to remove any residual material.

At this point in the photolithographic process, it is possible to check the quality of the image in the photoresist. The develop check verifies that the photoresist quality and the alignment are sufficient for the particular device. All wafers that pass this inspection step are ready for hard bake. The hard bake increases the adherence of the photoresist to the surface of the wafer, and evaporates more solvents. The considerations and types of equipment used for this operation are the same as for the soft bake operation. The temperature is generally higher than the soft bake, but the times are generally comparable.

The etch step is the next process in the sequence, and is the most critical. The most common method of etching is to immerse the wafers in an etching solution at a predetermined temperature. Prior experience and knowledge of the etch rate determines the etch time. If the unprotected area is successfully etched, the wafer is ready for the next step. Otherwise, the wafer is reimmersed in the etching solution to remove the remaining material. A list of materials commonly encountered in semiconductor processing, and the chemicals comprising their etch solution are given in Table 11-1.

TABLE 11-1: Etches for Materials in Semiconductor Processing

Materials	Etch
SiO_2	HF, NH_4F (buffered oxide etch)
aluminum	phosphoric acid, acetic acid, nitric acid
polycrystalline silicon	HF, HNO_3, acetic acid, KOH
silicon nitride	phosphoric acid

An etching technique that is gaining popularity is "plasma etching." Wafers masked with photoresist are placed in a chamber and evacuated, and a small amount of reactive gas is allowed back into the chamber. An electromagnetic field is applied, and the layer not protected by the photoresist is etched away. The technique holds much promise for the future, though there are still some problems with it. A plasma etching apparatus is shown in Figure 11-5.

Following the etch step, the photoresist must be removed before the final inspection can be performed. The photoresist may be removed by dissolving it in solvents, chemically removing it in hot acid baths, or reactively oxidizing it in using plasma techniques. The first two methods are most popular because of a history of proven success. Prior to proceeding to the next operation, the wafers undergo a final inspection. Wafers not meeting certain standards are sent back to be done again, or are removed from the line. All good wafers are sent on for subsequent processing.

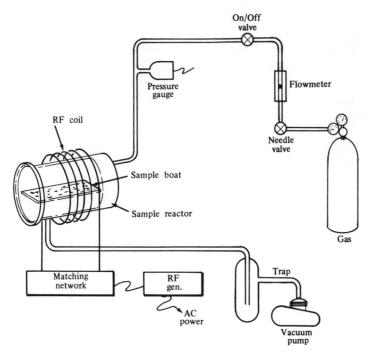

Figure 11-5 Plasma etching apparatus.

REVIEW EXERCISES: PHOTOMASKING

1. Following the develop step in a process using positive resist, is the resist left in areas protected by opaque regions of the mask?
2. Using Figure 11-3, determine:
   a. The spin speed needed to obtain a 1.6μ thick layer of AZ-1350J resist.
   b. The thickness resulting from a 6000 rpm application using AZ-111 resist.
3. Is a liquid that flows more slowly than another more or less viscous?
4. Name and describe two types of bake oven.
5. List four types of frequently used photomasks and give an advantage for each type.
6. Define photolithography.

7. List and define the four parameters that affect the performance of photoresist.

8. Why is priming necessary in some photoresist processes?

9. What is the most common method of applying photoresist during the manufacture of semiconductor devices?

10. What two parameters are used to control the quality of the finished photoresist layer?

11. What is the purpose of the develop check step?

12. Explain the difference between the soft bake and hard bake operations.

12

Chemical Vapor Deposition

12-1 INTRODUCTION

Chemical vapor deposition (CVD) is the formation of a stable compound on a heated substrate by the thermal reaction or decomposition of gaseous compounds. Epitaxial growth is a type of chemical vapor deposition, but a highly specific type. It requires that the crystal structure of the substrate be continued through the deposited layer. For this reason, epitaxial growth was covered in an earlier section. In this section, nonepitaxial CVD and its applications will be covered.

Chemical vapor deposition may be accomplished in many ways, but all types of CVD equipment need to have certain basic sections. These are:

1. Reaction chamber
2. Gas control section
3. Time and sequence control
4. Heat source for substrates
5. Effluent handling

The variety of ways of accomplishing each of these sections leads to a great number of individual reactor configurations.

The purpose of the *reaction chamber* is to provide a controlled envelope around the reaction zone. Systems are generally broken down into different types:

1. *Horizontal systems.* Wafers are placed horizontally on a wafer holder (boat or susceptor) as shown in Figure 12-1. In these systems, the gas flows in one end of the tube, across the wafers, and out the other end.

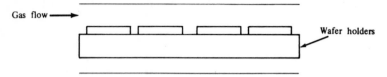

Figure 12-1   A horizontal reaction chamber

2. *Vertical systems.*   Wafers are placed on a susceptor with the gas flow incident to the wafers from the top as shown in Figure 12-2. The susceptor usually rotates to produce uniform temperatures.

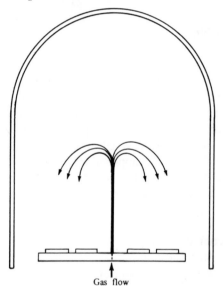

Figure 12-2   A vertical reaction chamber.

3. *Cylindrical or barrel systems.* Wafers are placed vertically on the outer surface (or sometimes the inner surface) of a cylinder. Gases flow into the chamber from the sides, and the susceptor usually rotates. Such a chamber is shown in Figure 12-3.

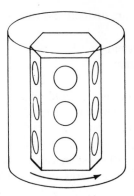

Figure 12-3   A cylindrical or barrel system.

4. *Gas-blanketed downflow system.* Gases flow downward as in a vertical system while wafers are on a moving wafer holder as in a horizontal system. A blanket of inert gas (usually nitrogen) keeps the reaction species separate from the outside atmosphere as shown in Figure 12-4.

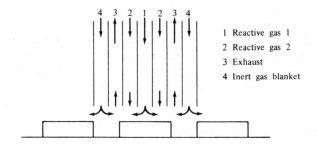

1 Reactive gas 1
2 Reactive gas 2
3 Exhaust
4 Inert gas blanket

Figure 12-4   A gas-blanketed downflow system.

The gas flow control section controls the amount of gas that flows into the reaction chamber. The exact type of flow controller used depends on the accuracy needed in the particular application. In general, the greater the percentage of control needed, the more important the flow controller becomes.

A time and sequence control section is responsible for the overall running of the CVD equipment. It may vary in complexity from manual on/off buttons controlled by an operator to a completely computer-controlled automatic programmer.

Heating sources are divided into two general categories:

1. Cold-wall system
2. Hot-wall system

This distinction is made because in cold-wall systems, reactions that lead to deposition on the chamber walls proceed at a relatively slow rate. In hot-wall systems, the deposition process will take place as fast or faster on the reaction chamber walls than on the wafers and the susceptor. Heating in a cold-wall CVD system can be accomplished through the use of RF (radio frequency) energy or UV (ultraviolet) energy. In an RF-heated susceptor, the energy in an RF coil is coupled into a coated carbon susceptor. The wafers are heated through their contact with the susceptor. Ultraviolet heating is accomplished by the use of light bulbs that emit strongly in the UV spectrum. The large amounts of energy from these bulbs heat the wafers and their holders by radiation. In both types of cold-wall heating, the walls of the chamber are only cold in comparison to the wafers themselves. Radiation and conduction from the susceptor produce a large temperature rise in the chamber walls. Hot-wall systems are heated using thermal resistance heating as in a diffusion furnace. In addition to the advantages of less deposition on the walls of a cold-wall system, the wafers may be heated and cooled much more rapidly because of the small thermal mass of the system and the relatively large gas-flow velocities.

The last section of a CVD system is the effluent handling section. All unreacted gas plus the carrier gas must be exhausted in some manner. Generally, the exhaust gases are cleaned of any harmful or reactive gases, cooled, and vented to the atmosphere.

12-2   CVD PROCEDURES AND USES

Chemical vapor deposition can be used to deposit many materials, but in semiconductor processing, the materials generally encountered in addition to epitaxial silicon are:

1. Polycrystalline silicon
2. Silicon dioxide (both doped and undoped)
3. Silicon nitride

Each of these materials may be deposited in a variety of ways, and each has many applications.

Polycrystalline silicon is silicon with a short-range crystal structure but no long-range crystal structure. It may be deposited if the deposition rate on a substrate is high, if the substrate has no crystal structure, or if the deposition temperature is below the threshold for single-crystal growth. Two general methods for deposition of polycrystalline or "poly" are:

Reaction	Carrier Gas	Deposition Temperature (°C)
$SiH_4 + Heat \rightarrow Si + 2H_2$	H_2	850–1,000
$SiH_4 + Heat \rightarrow Si + 2H_2$	N_2	600–700

The crystal structure of the poly depends on both the deposition temperature and the rate, and may be tailored for a particular application. Polycrystalline silicon is usually deposited undoped, and is doped later in the processing to provide a conductive layer for use in devices. The thickness of a polycrystalline layer may be determined by interference techniques.

Silicon dioxide may be obtained by using any of the following reactions:

Reaction	Carrier Gas	Deposition Temperature (°C)
$SiH_4 + Co \rightarrow SiO_2 + 2H_2$	H_2	600–900
$SiH_4 + 4CO_2 \rightarrow SiO_2 + 4CO + 2H_2O$	N_2	500–900
$2H_2 + SiCl_4 + CO_2 \rightarrow SiO_2 + 4HCl$	H_2	800–1,000
$SiH_4 + 2O_2 \rightarrow SiO_2 + 2H_2O$	N_2	200–500

Silicon dioxide may also be deposited containing arsenic, phosphorus, or boron by including some arsene, phosphene, or diborane in the reaction. These impurities form oxides that are readily incorporated into the layer of deposited silicon dioxide. Deposited oxide may be used as a predeposition source if it is doped, or as a barrier for masking; but its primary use is as a scratch protection layer over already completed circuits, and metallization. To avoid problems with the already deposited metallization, the deposition is generally performed at temperatures below 500°C. A multilayer structure of phosphorus-doped layer/undoped oxide layer or of undoped layer/phosphorus-doped layer/undoped oxide layer is often used as shown in Figure 12-5.

The multilayered structure is needed because:

1. The phosphorus-doped oxide is a chemical barrier to prevent the movement of contamination through the layer. It

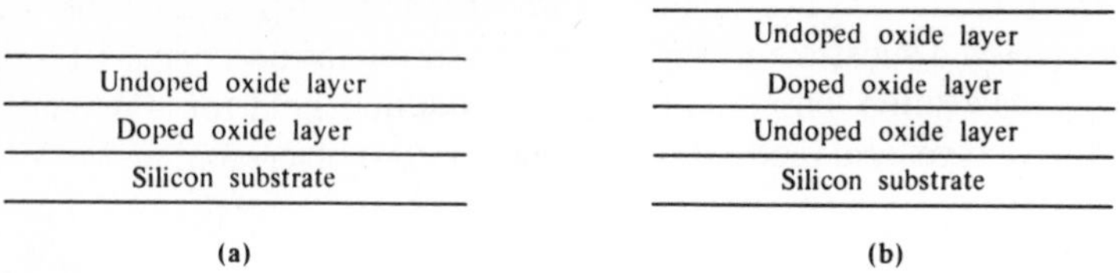

Undoped oxide layer
Doped oxide layer
Silicon substrate

(a)

Undoped oxide layer
Doped oxide layer
Undoped oxide layer
Silicon substrate

(b)

Figure 12-5   Cross section of silicon dioxide deposited for scratch protection;
(a) doped oxide capped with undoped oxide; **(b)** doped oxide with undoped
oxide both above and beneath.

also reacts with water to produce an electrolytic etching
action on the underlying aluminum metallization in the
presence of an applied voltage.

2. To prevent the etching action; an undoped layer of SiO_2
may be used either below, or above and below the phosphorus-doped layer of SiO_2.

The thickness of a deposited layer may be determined by looking at the
color chart for thermally grown SiO_2. Deposited SiO_2 is not as dense as
thermally grown SiO_2, but heating it to 900°C or higher for 30 minutes
results in properties that are almost indistinguishable.

The concentration of phosphorus in a layer of deposited SiO_2
is often determined by simultaneously depositing the layer in a lightly
doped p-type silicon wafer. The wafer is then diffused for a predetermined time in a furnace set to a standard temperature. A 4-point probe
reading is sufficient to determine whether the deposited SiO_2 contained
the desired amount of phosphorus. Figure 12-6 is a graph of the sheet
resistance of a wafer following a 30-minute predeposition at 1,000°C
versus the phosphorus concentration in the deposited oxide layer.

Silicon nitride is a dense dielectric often used to passivate circuits with device parameters sensitive to contamination, or for the controlled local oxidation of silicon. It can be deposited using CVD
techniques as follows:

Reaction	Carrier Gas	Deposition Temperature (°C)
$3SiH_4 + 4NH_3 \rightarrow Si_3N_4 + 12H_2$	H_2	900–1100
$3SiH_4 + 4NH_3 \rightarrow Si_3N_4 + 12H_2$	N_2	600–700

The thickness of deposited layer of silicon nitride (Si_3N_4) can
be determined fairly accurately through the use of a color chart, as SiO_2
thicknesses are determined. But, because the optical properties of Si_3N_4

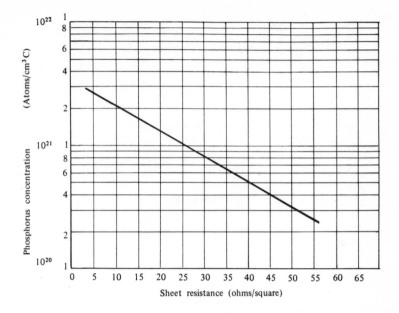

Figure 12-6 Phosphorus concentration in deposited oxide layer vs. resistance.

differ from those of SiO_2, a different relationship between the observed color and the film thickness exists. The relationship is shown in Table 12-1.

TABLE 12-1: Color Chart for Thermally Grown Si_3N_4 Films

(FILMS OBSERVED PERPENDICULARLY UNDER DAYLIGHT FLUORESCENT LIGHTING)

Film thickness		Color and comments
Å	μ	
380	.038	tan
530	.053	brown
750	.075	dark violet to red violet
900	.090	royal blue
1130	.113	light blue to metallic blue
1280	.128	metallic to very light yellow-green
1500	.150	light gold or yellow-slightly metallic
1650	.165	gold with slight yellow-orange
1880	.188	orange to melon
2030	.203	red-violet
2250	.225	blue to violet-blue

TABLE 12-1 (*cont.*)

Film thickness Å	μ	Color and comments
2330	.233	blue
2400	.240	blue to blue-green
2550	.255	light green
2630	.263	green to yellow-green
2700	.270	yellow-green
2780	.278	green-yellow
2930	.293	yellow
3070	.307	light orange
3150	.315	carnation pink
3300	.330	violet-red
3450	.345	red-violet
3530	.353	violet
3600	.360	blue violet
3680	.368	blue
3750	.375	blue-green
3900	.390	green (broad)
4050	.405	yellow-green
4200	.420	yellowish (not yellow but where yellow is expected)
4280	.428	light orange
4350	.435	light orange or yellow to pink borderline
4500	.450	carnation pink
4720	.472	violet-red
5100	.510	borderline between violet & blue-green; looks greyish
5400	.540	blue-green to green (quite broad)
5780	.578	yellowish
6000	.600	orange
8200	.820	salmon
8500	.850	dull light red-violet
8600	.86	violet
8700	.87	blue-violet
8900	.89	blue
9200	.92	blue-green
9500	.95	dull yellow-green
9700	.97	yellow to yellowish
9900	.99	orange
10000	1.00	carnation pink
10200	1.02	violet-red
10500	1.05	red-violet
10600	1.06	violet

REVIEW EXERCISES:
CHEMICAL VAPOR DEPOSITION

1. Briefly describe the difference between a hot-wall and a cold-wall CVD system.

2. Name three nonepitaxial materials that can be deposited using CVD techniques.

3. Following a 30-minute predeposition at 1,000°C, your test wafer has a sheet resistance of 35 Ω/square. Determine the phosphorus concentration in the layer of deposited SiO_2.

4. Explain the purpose of the reaction chamber in chemical vapor deposition.

5. List and describe the five main sections of a chemical vapor deposition system.

6. Explain the difference between epitaxial growth and chemical vapor deposition.

7. Give a reaction which may be used to deposit silicon nitride.

13

Metallization

After the devices in the silicon substrate have been fabricated, they must be connected together to perform circuit functions. This process is called metallization, and is performed using one of several available vacuum deposition techniques. In this section, we will look at the requirements of metallization systems, methods of depositing metals and other materials, and additional considerations in metallization.

13-1 METALLIZATION REQUIREMENTS

To serve as an effective interconnect metallization on silicon, the metal chosen must meet all of the following requirements to at least a satisfactory level:

1. Low-resistance electrical contact to the silicon.
2. Limited reactivity with silicon for a stable contact.
3. High electrical conductivity so high current is easily carried without voltage drops.
4. Good adherence to the underlying silicon dioxide or other dielectric.
5. A pattern must be easily definable in the layer.
6. The deposition method must be compatible with already existing structures.
7. The metallization must uniformly cover steps in the surface topography.

8. The metallization must be able to withstand "electromigration." (Electromigration is the migration of the atoms in the metallization caused by the flow of current.)

9. The metallization must not corrode under normal operating conditions.

10. It must be possible to bond easily to the metallization to allow external connection.

11. The metallization must be economically competitive.

No one metal perfectly meets all of these requirements. However, aluminum does meet all of these requirements quite well. Accordingly, aluminum is the metal most often chosen for device interconnection. Recent work has shown that the performance of aluminum can be improved upon by the introduction of small amounts of other elements. The tendency of aluminum to react with silicon can be halted by introducing a small percentage of silicon in the aluminum during deposition. In a similar fashion, the electromigration resistance of aluminum can be greatly increased by including a small percentage of copper in the deposited layer during deposition.

In instances where aluminum does not meet the requirements of the metallization, multilayered structures are often utilized. Each layer will meet some of the requirements, and a combination of the layers will result in satisfying all of them.

13-2 VACUUM DEPOSITION

Metallization is often applied through the use of a vacuum deposition technique. There are many types of systems, but they all have some characteristics in common. To perform any type of vacuum deposition, a system must have the following:

1. A chamber that can be evacuated to provide a sufficient vacuum for the deposition to take place. (This must include valves, etc. for the job.)

2. Vacuum pump (or pumps) to reduce the gases in the chamber to an acceptable level.

3. Instrumentation to monitor the vacuum level and other system parameters.

4. A method of depositing the wanted layer or layers of material.

Each of these needs can be met in many ways, but the trade-offs in-

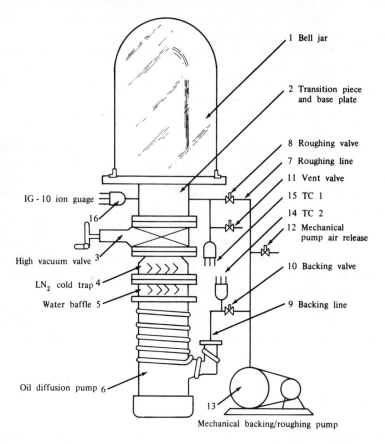

1 Bell jar

2 Transition piece
and base plate

8 Roughing valve

7 Roughing line

11 Vent valve

IG - 10 ion guage

15 TC 1

16

14 TC 2

12 Mechanical
pump air release

High vacuum valve 3

LN$_2$ cold trap 4

10 Backing valve

Water baffle 5

9 Backing line

Oil diffusion pump 6

13

Mechanical backing/roughing pump

Figure 13-1 Schematic diagram of typical fast-cycling, high-throughput vacuum coating system.

volved must be considered. A typical vacuum deposition system is shown in Figure 13-1.

A vacuum chamber consists of a leak-free enclosure allowing sufficient access for work and instrumentation. Both glass and stainless steel enclosures are common, but stainless steel accommodates more nonstandard configurations and does not break.

To obtain sufficient vacuum, different types of pumps are used. Pumps are like gears on a car—different types work better over different vacuum ranges. A summary of different types of pumps and the pressure ranges over which they are used is given below:

1. Atmospheric pressure to intermediate vacuum levels (10–100μ)

   a. *Rotary oil-sealed pumps*. This type of pump uses a rotor that is sealed against leaking by a vacuum oil. The air left in the vacuum system enters the pump through the inlet port, is compressed, and ejected to the atmosphere through the exhaust or discharge port.

   b. *Sorption pump*. This type of pump uses chemicals that will adsorb gases on their surface. Containers of these chemicals adsorb gases until no more can be accommodated (usually many cycles) and then must be baked out to restore this capacity.

2. Intermediate vacuum levels to low vacuum levels (25μ–10^{-6}mm)

   a. *Diffusion pump*. In this pump, vapor from a boiler passes through a series of nozzles in a downward direction, carrying residual atoms in the vacuum chambers with it.

   b. *Turbomolecular pump*. This pump has a series of blades set around a hub, many levels deep (similar to an electrical turbine) to propel molecules out of the chamber by imparting suitable momentum to them.

3. Low vacuum levels to ultralow vacuum levels (10^{-6} T. 10^{-10})

   a. *Ion pump*. Using a combination of an electric and a magnetic field, this pump provides a method of ionizing atoms and then trapping the ions.

The instrumentation necessary for a vacuum deposition system must provide:

1. A method of determining the vacuum level in the chamber.
2. A method of measuring the status of all valves, etc., in the system.
3. A method of determining the thickness of any deposited layers.

The vacuum level inside the vacuum chamber can be determined from atmospheric pressure to intermediate vacuum levels using a diaphragm that moves with changes in the pressure. A mechanical or electrical readout can be used. For better vacuums, a measure of the ability of the residual gas to carry heat away from a filament is often used. The more gas present, the more heat can be carried away. This principle is used in both the Termocouple and the Pirani Gauge.

13-3 DEPOSITION TECHNIQUES

There are five methods of depositing materials using vacuum techniques in the semiconductor industry. These methods are:

1. Filament evaporation
2. Electron-beam evaporation (E-beam)
3. Flash evaporation
4. Induction evaporation
5. Sputtering

Each of these methods has advantages and disadvantages, and the trade-offs involved must be considered when selecting the deposition method.

Filament evaporation is the simplest and least expensive deposition method. The evaporation takes place from a filament or a boat heated by thermal resistance heating. Figure 13-2 shows a typical filament evaporation system.

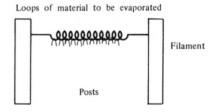

Figure 13-2 A typical filament evaporation system.

The evaporation is accomplished by gradually increasing the current flowing through the filament to first melt the loops of material thereby wetting the filament. (Care must be taken to choose a filament compatible with the material to be evaporated.) Once the filament is wetted, the current through the filament is increased to accomplish the evaporation. Filament evaporation systems are easily set up and many materials can be evaporated using them. However, the contamination level of the deposited materials is often sufficiently high to interfere with the functioning of the device. The contamination may come from the filament, or from poor handling techniques. For this reason, filament evaporation of aluminum is not common. Care can be taken to minimize this effect, but other methods have proved more economical and reliable. Filament evaporation is often used to deposit backside gold, however, since contamination is not of concern in this case. This technique cannot be used to evaporate composite materials, because the element with the lowest melting point evaporates first, leaving the rest of the material to evaporate later.

Electron-beam evaporation (frequently called E-beam) uses a focused beam of electrons to heat the material. A high-intensity beam of electrons is generated in a manner similar to that used in a television picture tube. The focused beam of electrons melts the material contained in a water-cooled block with a large depression called a hearth. Because only electrons come in contact with the material to be evaporated, it can be a low-contamination process. It is also a rapid process, but cannot be used for the deposition of composite materials unless more than one hearth is employed. Because of the intense electron-beam source used, there is often radiation damage to the substrates the material covers, which must be annealed out later in the process. A typical electron-beam evaporation system is shown in Figure 13-3.

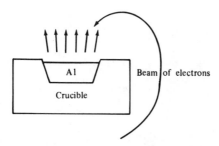

Figure 13-3   A typical electron-beam evaporation system.

Flash evaporation is similar to filament evaporation, in that the material is evaporated by thermal resistance heating, but the similarity ends there. Flash evaporation uses a continuously fed spool of wire (or in some cases, stream of pellets or powder) incident on a heated ceramic bar for the deposition as shown in Figure 13-4. This deposition technique combines the speed and contamination-free features of E-beam deposition with the radiation-free feature of filament evaporation, and still offers the option of depositing composite layers.

Induction evaporation is a recent innovation that offers some attractive features for certain coating problems. A radio-frequency source such as the one used in epitaxial deposition is used to couple

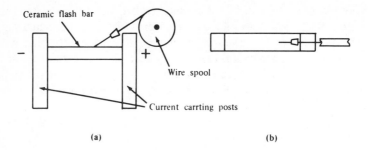

Figure 13-4 A flash evaporation system; **(a)** side view; **(b)** top view.

power into the metal to be evaporated in the crucible. The energy melts the metal, resulting in evaporation from certain regions as shown in Figure 13-5. This method is not in common use in semiconductor operations.

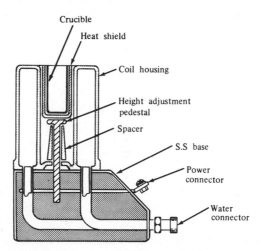

Figure 13-5 An induction evaporation source.

Sputtering is the last vacuum deposition method encountered in semiconductor processing. In sputtering, ions of inert gas are introduced into the chamber after a satisfactory vacuum level has been reached. An electric field ionizes these atoms, and causes them to move to one plate in the chamber called the target. When the ions strike the target, they dislodge atoms from it, depositing them on the substrates facing the target as shown in Figure 13-6. Sputtering can be accom-

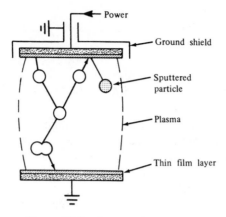

Figure 13-6   A sputtering system.

plished using both DC and RF voltages, and it can be used to deposit almost any material, although the deposition rate is often extremely low. Adhesion of layers deposited using this technique is generally very good.

13-4   VACUUM DEPOSITION CYCLE

The following metallization sequence is typical, regardless of the particular material being deposited or type of equipment being used in a given manufacturing facility:

1. Remove all contamination, etc., from the wafers and dry them.
2. Position the wafers in the vacuum chamber to receive a uniform layer. (In many cases, a rotating structure called a planetary is used. It also guarantees that the coverage of steps on the surface is as good as possible.)
3. Close vacuum chamber and "rough" down to ~25μ.
4. Close the valve to the roughing pump and open the valve to the high vacuum pump and pump until the required vacuum is reached (10^{-6}–10^{-7}mm is typical.)
5. Turn on the source and evaporate a small amount of material into a shield between the source and the wafers to clean the source.
6. Deposit the necessary thickness of material on the substrates (the substrates may be heated to increase adhesion).

7. Turn the source off and cool.

8. Fill the chamber with an inert gas like N_2 and then open it.

REVIEW EXERCISES: METALLIZATION

1. List five requirements that a metallization must meet.
2. List and briefly describe three vacuum deposition techniques.
3. Which vacuum deposition technique may lead to radiation damage?
4. What is the name of the rotating structure that wafers are mounted on during a vacuum deposition?
5. Why is aluminum the most frequently used metal for the process of metallization?
6. Why are trace amounts of silicon and copper added to the aluminum during metallization?
7. Name the major components of a vacuum deposition system.
8. List and explain four methods of depositing metals using vacuum techniques.
9. Describe a typical vacuum deposition cycle.

14

Device Processing: from Alloy to Sale

The remaining processing steps that devices undergo between metallization and final sale are as important as the initial steps. However, the "back end" of the line does not have the glamour of the rest of processing. But, as the price of silicon chips continues to fall, the companies that package, test, and distribute the devices most efficiently will remain strong in the marketplace. The device flow for the remainder of a processing line is discussed below:

14-1 ALLOY/ANNEAL

The successful etching of the aluminum on the front side of the wafer to form the device interconnection does not guarantee that a good electrical contact has been formed. A subsequent "alloy" step is usually used to insure low-resistance contact between the aluminum and the silicon. The alloy step is performed in a diffusion furnace set at a relatively low temperature. The alloy temperature and time will vary from process to process, but the limits on the temperature can be determined in part by looking at the diagram of the aluminum–silicon system in Figure 14-1.

The line indicated by the arrows shows the lowest temperature at which a completely molten mixture exists. This temperature varies with the atomic percent of silicon in the aluminum as indicated by the figure. The lowest temperature at which a molten solution exists is the one at the intersection of the two lines at 577°C. This temperature is

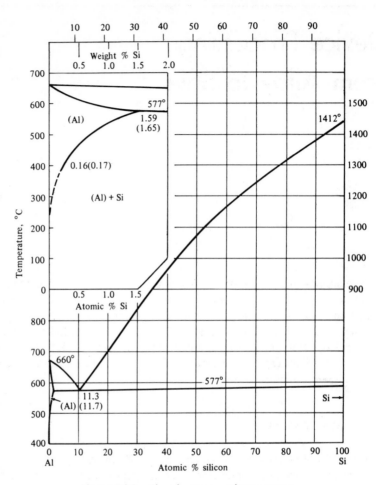

Figure 14-1   The aluminum–silicon system.

the aluminum–silicon eutectic temperature. If an aluminum–silicon mixture is heated to above 577°C, melting will occur, ruining any devices that are present. The upper limit in the alloy temperature is thus 577°C. The lower temperature is set by process considerations like cleanliness and the aluminum deposition temperature. Most alloy steps are performed at temperatures between 450°C and 550°C for times of between 10 and 30 minutes.

Either during the alloy step or directly following it, the wafers are often exposed to a gas mixture containing hydrogen (or occasionally another gas). This step is usually called an "anneal" step. The anneal step is designed to optimize and stabilize device characteristics. Hydrogen is thought to combine with uncommitted atoms at or near the silicon–silicon dioxide interface, thus reducing their effect on device performance. Typical anneal temperatures are 400°–500°C for times of 30 minutes to 60 minutes.

14-2   POST-ALLOY SAMPLE PROBE

Following the metallization, etching, alloying, and annealling steps, the wafers should contain fully functional devices. If the wafers contain only discrete devices such as diodes or transistors, it is easy to test a certain fraction of the devices to see if they function properly. If more complicated integrated circuits have been fabricated, it is usually necessary to test some of the diodes, resistors, transistors, etc., that must all work for the circuit to function properly. In this manner, it is possible to discard wafers that have no chance of having an economical number of good die on them. However, this probing step provides more than just a quick check on the wafers from the fabrication line. By properly selecting the devices that are tested, and by taking data beyond a pass or fail level, it is possible to trace variations that are occurring in the fabrication process. For instance, variations in the value of a base resistor may indicate a change in the amount of boron in the base of a transistor. This change may lead, in turn, to a change in the gain of the transistors that are being fabricated. By anticipating variations in transistor gain, it may be possible to correct for a small problem before it becomes a major one.

Post-alloy sample probe is often performed using an oscilloscope and a hand probe station. On a typical wafer, the performance of devices from different areas of the wafer will be measured as shown in Figure 14-2. Measurements taken using this or a similar pattern will provide information on processing variations that exist across a wafer.

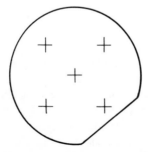

Figure 14-2   Sample areas on a typical wafer.

14-3   SCRATCH PROTECTION

To protect the devices on the wafers from improper handling and chemical contamination, a layer of deposited oxide is often used. The deposition considerations for this layer were discussed in the chemical vapor deposition section. Following the deposition step, openings in the layer of deposited oxide are etched over the bonding pad areas. The wafers are now ready for any backside preparation necessary prior to wafer sorting.

14-4   BACKSIDE PREPARATION

The backside of a wafer may have to be altered to prepare for subsequent processing steps. Two types of preparation that are common are:

1. *Backside lapping.* The backside of a wafer may be lapped to remove diffused layers that interfere with the electrical properties, to thin the wafer to make it easier to separate the die, or to prepare the backside for a subsequent metal deposition.
2. *Backside metal deposition.* A metal such as gold may be deposited on the wafer back to make the attachment of the separated die to the package easier in a later operation.

Neither of these steps, one of these steps, or both of these steps may be used on wafers to prepare them for later operation.

When a metal is used for backside contact, it is usually deposited using filament evaporation. Gold is often chosen for the backside metal because of the low gold–silicon eutectic temperature. We can see from Figure 14-3 that this temperature is 370°C. It is sufficiently low to prevent degradation of other device characteristics when the gold is alloyed to the silicon.

14-5   WAFER SORT

The wafers are now at their moment of truth—do they contain any functional die? To determine this fact, the wafers are placed on a wafer prober (usually computer controlled) and the individual die on each wafer is tested. Pointed metal probes contact each bonding pad and supply the necessary currents and voltages to the device. The devices that function properly are left alone, while those that fail are marked

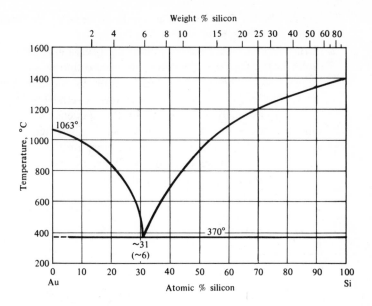

Figure 14-3   The gold–silicon system.

with a drop of ink from an "inker." In some instances, these may be differentiated as top-grade devices and ordinary-grade devices. In this case, two different types of ink may be employed to differentiate between the grades. The ink is usually baked to prevent running at any later step.

14-6   DEVICE SEPARATION

With the devices or circuits on the wafer tested, it is time to separate them into individual die for final packaging. This operation is generally called "wafer scribe," although changes in the methods used to accomplish the separation have made this term technically obsolete. The three methods commonly used to separate die are listed and described below:

    1. *Diamond scribing.* A tool with a precisely shaped diamond imbedded in the tip is drawn across the wafer along the scribe line, making a "mark" or "scribe" in the wafer. The imperfection in the crystal structure caused by the scribing defines the crystal planes along which the wafers prefer to break. By bending the wafer on both sides of the scribe line, the wafer is broken along the line.

2. *Laser scribing.* A laser is pulsed to generate a series of holes in the silicon wafer along the scribe line as shown in Figure 14-4. The series of holes serves as the line along which the wafer breaks. This technique is relatively recent, and is occasionally complicated by the condensation of the silicon initially evaporated by the laser. (The term often used is "kerf.") Backside laser scribing, or the use of a protective layer of material are two ways of preventing this kerf from impacting the device yield.

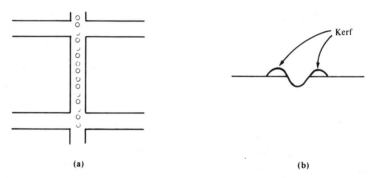

(a)   (b)

Figure 14-4 The pattern left by a laser scriber; **(a)** top view; **(b)** side view.

3. *Sawing.* An even more recent development than the laser scribing is the use of rotating blades to separate the die. Recent metallization advances have made it possible to manufacture saw blades capable of separating die with a minimum of silicon loss. Use of this method provides die that have uniform dimensions with square sides. These features make the process attractive for automated wafer-handling techniques.

14-7 DIE-ATTACH (OR DIE BONDING)

For convenience of handling, the good die are attached to some sort of a package. Three types of die attach are encountered: "eutectic" die-attach, "preform" die-attach, and "epoxy" die-attach. These methods of attaching a die to a wafer are discussed below.

Eutectic die-attach involves the prior deposition of a layer of metal such as gold on the backside of the die. By heating the package to above the eutectic temperature (370°C for gold–silicon) and placing the die on it, a bond is formed between the die and package.

Preform bonding involves the use of a small piece of special composition material that will adhere to both the die and package. A preform is placed on the die-attach area of a package and allowed to melt. The die is then "scrubbed" across the region until the die is attached, and then the package is cooled.

Epoxy bonding involves the use of an epoxy glue to attach the die to the package. A drop of epoxy is dispensed on the package, and the die is placed on top of it. The package may need to be baked at an elevated temperature to cure the epoxy properly. Epoxy die-attach can be performed using either electronically conductive or nonconductive material, while the other two die-attach methods form conductive bonds.

14-8   WIRE BONDING

The type of wire used to connect the device externally is usually aluminum or gold. Gold is considerably more expensive than aluminum, but offers the advantages of corrosion resistance and higher current-carrying capability. Two methods of attaching the wires between the device pads and the packages are thermal compression (TC) bonding and ultrasonic (US) bonding. In each case, the name of the bonding method is an accurate description of the steps involved. Thermal compression bonding is often used with gold wire and involves heating the package and forming the bond between the wire and the pad using both heat and pressure. Ultrasonic bonding uses a pulse of ultrasonic energy to provide a scrubbing action that forms a bond between the wire and the pad. Ultrasonic bonding is usually used with aluminum wire.

14-9   PACKAGE CONSIDERATIONS

Semiconductor processing technology has evolved to the point that the cost of the package may be a considerable fraction of the total cost of the device. For this reason, packaging considerations recently have drawn much attention. The prime consideration is the material used to construct the package. The oldest and generally most reliable package is either metal or metal and ceramic. These packages tend to be the most expensive, so replacements for them are constantly being searched for. The use of various plastic and epoxy packages has become popular in recent years because of their low cost and the ease of forming the package. The results with these packages are better every year, but they still cannot match the metal or metal and ceramic package.

Another consideration that must sometimes be made is a package's ability to conduct excess heat away from the die. Special metal tabs, fins, or wings may be designed as part of the package to conduct heat away from the die while it is in operation.

14-10 FINAL TEST

Once a device is packaged, it is ready for final test. The test the device undergoes may well be the one it underwent after wafer sort, but the handling steps involved in the bonding and packaging may have damaged the die, or these steps may not have been done correctly. If either has occurred, the packaged device will not perform properly and hence this test is necessary.

14-11 MARK AND PACK

Once the bad devices have been removed from the rest of the packaged devices, the last step prior to storing them is called mark and pack. The packages are marked with the device code and date code that tells customers when they were manufactured. The devices are now ready to sell and ship to waiting customers.

REVIEW EXERCISES: DEVICE PROCESSING: FROM ALLOY TO SALE

1. What two compositions of gold–silicon are just at the melting point at 800°C?
2. What is the eutectic composition of aluminum–silicon?
3. Is the gold–silicon or the aluminum–silicon eutectic temperature higher?
4. What are two methods of separating die?
5. State two methods of connecting wire leads from a device to a package.

6. What is the purpose of the post-alloy probe step?
7. Explain the two types of backside wafer preparation.
8. Why is gold frequently used as the backside metal?
9. How are non-functional die determined and identified?
10. List the various steps utilized during the post-alloy to shipment of a completed semiconductor device.

15

Devices

Fabrication technologies can be divided roughly into two categories:

1. Bipolar Technology
2. Metal Oxide Semiconductor (MOS) Technology

Although bipolar and MOS technologies are both based on the same basic processing steps, the processing sequence and the surface geometrics used, produce transistors that function on different physical principles.

15-1 BIPOLAR TECHNOLOGY

The word "bipolar" is derived from the flow of both holes and electrons in the functioning of the transistor. A typical bipolar sequence consists of seven or more masking steps. The sequence used as the basis of a bipolar process is:

[Mask 1.] 1. Buried layer—the heavily doped n^+ region beneath the majority of all active devices.
2. Epitaxy—the n-type layer in which all devices are fabricated.
[Mask 2.] 3. Isolation—the p-type diffused region that provides electrical isolation between adjacent regions.

[Mask 3.] **4.** Base—the *p*-type diffusion that serves as the base of all *npn* transistors, and the body of most resistors.

[Mask 4.] **5.** Emitter—the n^+ diffusion that forms the emitter of *npn* transistors.

[Mask 5.] **6.** Contact—openings to provide electrical access to all devices.

[Mask 6.] **7.** Metallization—the conductive paths that electrically connect the devices to form a circuit.

[Mask 7.] **8.** Scratch protection—the deposited layer of SiO_2 that serves as both a physical and a chemical protective barrier over the completed circuit.

A cross section of a typical bipolar sequence is shown in Figure 15-1 on page 140.

15-2 DEVICES FABRICATED USING STANDARD BIPOLAR TECHNOLOGY

1. *npn* Transistors:

The bipolar *npn* transistor is the device the bipolar process is optimized to fabricate. These transistors are used as both amplifiers and switches in circuit designs. The current gain (also called h_{fe} or β) is the ratio of the current that flows in the collector to the current that flows in the base. The symbol for an *npn* transistor is shown in Figure 15-2. Figure 15-3 shows top and side views of a typical *npn* transistor. The current the transistor can handle is determined by the size of the device. The typical minimum geometry transistor can handle 1–10 mA. The base/collector reverse breakdown is determined by the doping on both sides of the junction depth. Current gain is typically 50–500.

2. *pnp* Transistors:

The symbol for a *pnp* transistor is shown in Figure 15-4.

A. Lateral *pnp* Transistor

This current gain of a lateral *pnp* transistor is typically less than that of a vertical *npn* transistor. Its current gain goes to one (i.e., no gain) at a much lower frequency than the current gain of a vertical *npn* transistor. Figure 15-5 shows top and side views of a lateral *pnp* transistor.

B. Vertical *pnp* transistor

A vertical *pnp* transistor can be used if the collector of the device in the circuit goes to the circuit ground (the substrate).

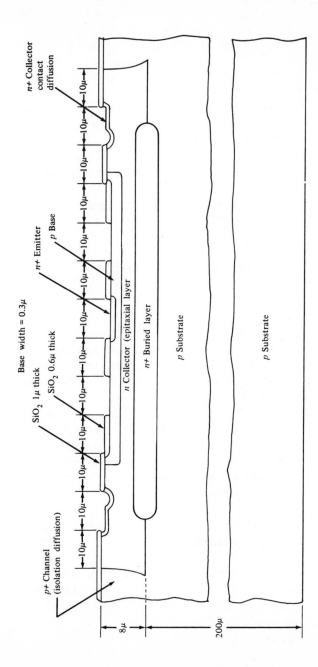

Figure 15-1 Cross section of a typical bipolar integrated circuit.

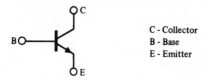

Figure 15-2   Symbol for a *npn* transistor.

C - Collector
B - Base
E - Emitter

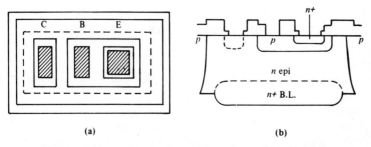

(a)

(b)

Figure 15-3   A *npn* transistor; (a) top view; (b) side view.

Figure 15-4   Symbol for a *pnp* transistor.

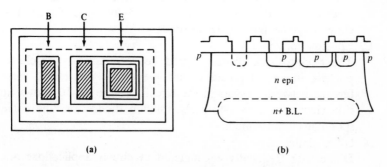

(a)

(b)

Figure 15-5   Lateral *pnp* transistor; (a) top view; (b) side view.

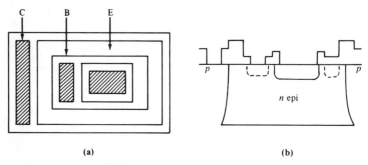

Figure 15-6 Vertical *pnp* transistor; (a) top view; (b) side view.

Figure 15-6 shows top and side views of a vertical *pnp* transistor.

3. Diodes:

A diode is present at every *pn* junction, but only a few of these diodes are used in a circuit application. A diode allows current to flow in the direction of the arrow, but does not allow any current to flow in the reverse direction until the breakdown voltage is reached. Figure 15-7 shows the symbols for both standard diodes and zener diodes.

A. Emitter/base diode:

This device is usually formed by shorting the collector and base of a *npn* transistor together as the anode, with the emitter as the cathode. This diode has a low (6–10 volt) reverse breakdown, and is often used as a zener diode. An emitter/base diode is shown in Figure 15-8.

B. Base/collector diode:

This diode is usually formed using the base of a *npn* transistor to form the anode and the collector to form the cathode. Typical reverse breakdown of this device is 15–50 volts. Figure 15-9 shows a typical base/collector diode.

C. Epi-isolation diode:

The back-to-back diodes formed between any two "isolated" pockets in an integrated circuit prevent any electrical interference between devices in the circuit. Figure 15-10 shows the side view of the back-to-back isolation diodes and the corresponding circuit symbol.

D. Other diodes:

Diodes not frequently encountered in circuit applications are:

Figure 15-7   Diode symbols; **(a)** standard diode; **(b)** zener diode.

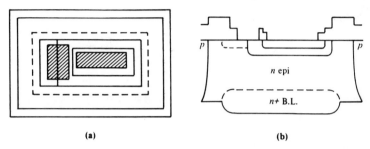

(a)   (b)

Figure 15-8   Emitter base diode; **(a)** top view; **(b)** side view.

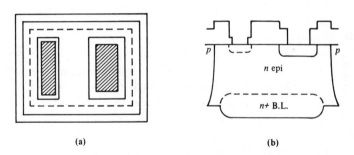

(a)   (b)

Figure 15-9   Base collector diode; **(a)** top view; **(b)** side view.

(a)   (b)

Figure 15-10   Isolation diodes; **(a)** side view; **(b)** circuit symbol.

    1. Emitter/isolation diode
    2. Isolation/buried layer diode

4. Resistors:
The symbols for resistors are shown in Figure 15-11. The current through a resistor is related to the voltage across it by the relationship:

$$V = RI$$

       (a)   (b)

Figure 15-11   Resistor symbols; (a) standard symbol; (b) pinched resistor.

    A. Base resistor:
    This resistor is made by contacting both ends of a region of p-type base diffusion. Typical resistor range is 50–50,000 Ω, since base sheet resistivities are generally in the range of 100–500 $\Omega/\square$. A typical base resistor is shown in Figure 15-12.
    B. Pinched-base resistor:
    This resistor is fabricated by diffusing an emitter region over the center of a base resistor. Typical sheet resistance values are 2000–10,000 $\Omega/\square$, but the control on the resistance is quite bad. Typical resistor values range from 10,000–500,000 Ω. Figure 15-13 shows a typical pinched-base resistor.
    C. Emitter resistor:
    A diffused emitter region is contacted at both ends to fabricate this resistor. To minimize area while preventing unwanted parasitic device action and retaining resistor control, the emitter diffusion is generally performed in a region that has previously received a base diffusion, and one end of the resistor is connected to the base region. Typical emitter sheet resistivities are 4–10 $\Omega/\square$, and emitter resistors of values 5–100 Ω are often made. A typical emitter resistor is shown in Figure 15-14.
    D. Epi resistor:
    This resistor consists of an epitaxial region of silicon surrounded by an isolation wall with n^+ contacts at each end. Typical sheet resistivities for epi vary from 400–2000 $\Omega/\square$. Control of epi resistor values is not as good as the control of diffused resistors,

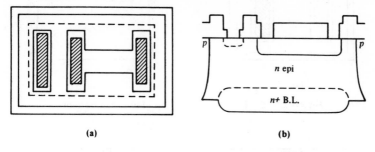

Figure 15-12 Base resistor; **(a)** top view; **(b)** side view.

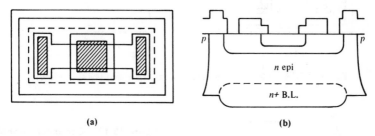

Figure 15-13 Pinched-base resistor; **(a)** top view; **(b)** side view.

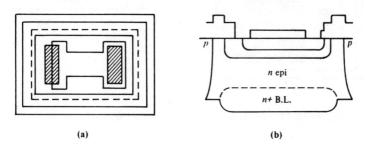

Figure 15-14 Emitter resistor; **(a)** top view; **(b)** side view.

because of the variation in epi thickness and resistivity, and variations in the lateral diffusion of the isolation. Typical epi resistor values are 1000–50,000 Ω. Figure 15-15 contains top and side views of an epi resistor.

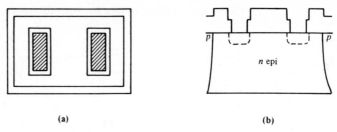

(a)   (b)

Figure 15-15   Epi resistor; **(a)** top view; **(b)** side view.

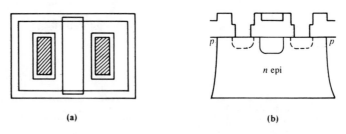

(a)   (b)

Figure 15-16   Pinched-epi resistor; **(a)** top view; **(b)** side view.

Figure 15-17   Capacitor symbol.

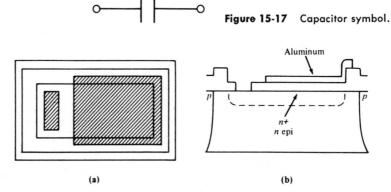

(a)   (b)

Figure 15-18   Dielectric capacitor; **(a)** top view; **(b)** side view.

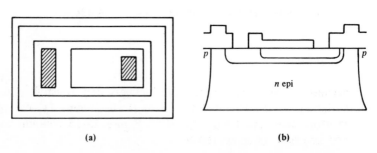

(a)   (b)

Figure 15-19   Junction capacitor; **(a)** top view; **(b)** side view.

E. Pinched-epi resistor:

This resistor is similar to an epi resistor, but has an additional *p*-type base diffusion over the center of the resistor, reducing the current-carrying region of the resistor, and hence increasing the sheet resistance. For this reason, pinched-epi resistors are often used instead of epi resistors in a design layout. A typical pinched-epi resistor is shown in Figure 15-16.

5. Capacitors:

Capacitors are used in circuits for charge-storage purposes or to supress circuit transients. Figure 15-17 depicts the circuit symbol of a capacitor.

A. Dielectric capacitors:

A capacitor is formed whenever a dielectric layer separates two conductive regions. The top plate of the capacitors is most often the interconnect metallization with one of the diffused regions (isolation, base, or emitter) used as the other plate of the capacitor. The intervening layer of thermal oxide is the dielectric. Capacitance per unit area increases with thinner oxide, so the thin emitter oxide is often used for capacitors. Figure 15-18 shows top and side views of a dielectric capacitor with the emitter diffusion as one plate of the capacitor.

B. Junction capacitors:

A reverse biased on junction behaves like a capacitor for small voltage excursions around the operating point. This type of capacitor is often used where the low leakage and constant capacitance of a dielectric capacitor are not needed. Figure 15-19 shows top and side views of a junction capacitor.

15-3 MOS TECHNOLOGY

Many variations of MOS technology exist, but the basic considerations are the same regardless of the particular steps selected for the fabrication process of metal oxide semiconductors. These steps may or may not coincide with the actual mask numbers. There may be as many as 10 MOS masking operations in these 5 steps (The minimum number of bipolar masks is typically 7.):

1. *Source-drain.* A *p*-type diffusion to form the resistors and the two current carrying terminals of the transistor.

2. *Gate oxidation.* The thin, carefully grown SiO_2 layer that the controlling charge acts through.

3. *Contact.* Openings to provide electrical access to the devices.

4. *Metallization.* The conductive paths that electrically connect the devices to form a circuit.

5. *Scratch protection.* The deposited layer of SiO_2 that serves as both a physical and a chemical protective layer over the completed circuit.

A cross section of a typical MOS circuit is shown in Figure 20.

The devices that can be fabricated using this technology include:

1. *MOS transistors.* The process is optimized to fabricate p-channel MOS transistors (Figure 15-21). This device can provide amplification, but is generally used as only an on/off switch in circuits.

2. *Source/drain resistor.* This is the only resistor available using the MOS process (see Figure 15-22). Typical ranges are 50–10,000 Ω, since source/drain sheet resistances are 50–200 $\Omega/\square$.

3. *Capacitor.* A capacitor is formed wherever a layer of SiO_2 covers a conductive region of silicon. The capacitance per unit area increases with thinner oxide layers, so the gate oxide layer is often used for capacitors. A MOS dielectric capacitor is shown in Figure 15-23.

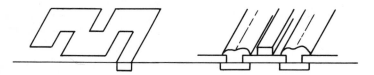

Figure 15-20   Cross section of a typical MOS circuit.

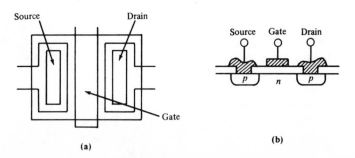

(a)

(b)

Figure 15-21   A p-channel MOS transistor; **(a)** top view; **(b)** side view.

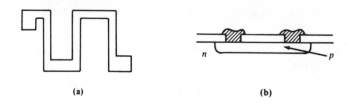

(a)   (b)

Figure 15-22   Source/drain resistor; **(a)** top view; **(b)** side view.

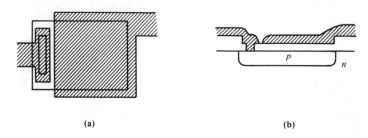

(a)   (b)

Figure 15-23   MOS dielectric capacitor; **(a)** top view; **(b)** side view.

15-4   OTHER MOS TECHNOLOGIES

Varieties in the basic *p*-channel MOS process include:

1. *n-channel MOS.* Start with a substrate doped with the opposite conductivity type (*p*-type) and diffuse *n*-type impurities.
   This technology offers faster devices than *p*-channel. It is
   called NMOS technology.
2. *Silicon gate.* Substitute a conductive layer of polycrystalline
   silicon for the metal as the gate. This layer can also be used
   as interconnect elsewhere in the circuit, in addition to the
   metal layer. Silicon gate technology results in more and
   faster devices per unit area. It is often referred to as SIGFET
   technology.
3. *CMOS.* Combine both *p*-channel and *n*-channel devices on
   the same chip by adding processing steps. Devices can be
   fabricated that use very little power to operate. CMOS
   stands for Complementary MOS.
4. *SOS.* This layer of silicon dioxide is deposited on an insulating sapphire substrate. The devices are fabricated in
   this thin epitaxial silicon layer. The letters stand for *S*ilicon-
   *O*n-*S*apphire.

REVIEW EXERCISES: DEVICES

1. What are the two categories that fabrication technologies can be divided into?
2. How do the number of masking steps in bipolar and MOS technology compare?
3. How can a diode be formed from a bipolar *npn* transistor?
4. Does a base resistor or a pinched-base resistor have a higher resistance per unit area?
5. Draw the cross-sectional view of a *p*-channel MOS transistor.
6. What is the purpose of the buried layer of the *npn* transistor in the bipolar process?
7. What is the purpose of the gate oxide in the MOS transistor structure?
8. What are the advantages of the dielectric capacitor in comparison to the junction capacitor?
9. What devices can be fabricated using MOS technology?
10. What devices can be fabricated using bipolar technology?

16

Contamination Control

Throughout the entire semiconductor fabrication process, it is critical to minimize the amount of contamination that comes in contact with the wafers and the wafer processing equipment. In this section, we look at the steps that are taken to achieve this result by considering the five general topics listed below. These topics are:

1. Chemicals and cleaning procedures
2. Water and rinsing procedures
3. Air
4. Gases
5. Personnel/clean room

Each of these topics will be considered separately for simplification, but in a working environment, they must all be considered simultaneously.

16-1 CHEMICALS AND CLEANING PROCEDURES

One of the first considerations that must be made in wafer processing is on how wafers can be cleaned prior to any process step. Attempts are made to keep wafers clean at all times, but prior to high-temperature processing steps such as diffusion, epitaxial growth, or chemical vapor deposition, even more care must be taken. The two types of contamination that have been found to cause the largest problems in semiconductors are ions that are mobile in silicon dioxide, e.g., sodium and elements that diffuse in silicon and precipitate out somewhere in the interior, e.g., gold and some other metals.

Sodium interferes with the normal operation of semiconductor devices by rapidly drifting through silicon dioxide toward regions with a negative bias. It then gives rise to changes in device characteristics such as excessive leakage. Sodium can be kept out of a fabrication line by specifying low-sodium chemicals and by rigorously enforcing proper wafer handling techniques. Sodium is a chemical present in the human body, and careless procedures will result in unwanted sodium contamination of wafers or wafer-handling equipment.

Certain elements are soluble in silicon at elevated temperatures, but precipitate into nonlattice locations when the wafer temperature is lowered. These elements interfere with the normal flow of holes and electrons in the silicon crystal when the device is in operation. Once a quantity of any of these elements has contaminated a wafer, it is impossible to completely remove it. However, proper cleaning prior to a high temperature operation minimizes its effect. Numerous methods of cleaning are popular in the semiconductor industry, but they all have certain common characteristics. The first step in dealing with wafers with an unknown history is to thoroughly degrease them. A common method is the use of a degreasing chemical such as 1,1,1-trichloroethane followed by rinses in acetone and alcohol. This cleaning procedure guarantees that any greases or waxes that might be insoluble in subsequent cleaning steps are removed. (If the history of the wafer is known, this degreasing operation may often be safely omitted.) Wafers are then sent through a series of solutions designed to remove any trace of metals or other potentially harmful materials. A common series of cleaning steps is:

STEP	REASONS
1. Heat in H_2SO_4	1. Removes any photoresist or other organic material
2. Heat in aqua regia	2. Dissolves gold as well as other metals
3. Dip briefly in dilute HF	3. Top layer of SiO_2 containing any potential contamination is etched away
4. Rinse in water	4. Remove any residual acid
5. Dry	5. Get them ready for the next process step

The chemicals used to remove trace amounts of harmful elements must themselves contain sufficiently low amounts of each of these elements.

In addition to cleaning the wafers that are to be processed, all fabrication equipment that comes in contact with the wafers, or is directly linked to wafer processing must be similarly cleaned. This includes diffusion tubes and associated glassware, wafer boats, push rods, thermocouple shields, vacuum wands, and many other items.

16-2 WATER

Water is used in the rinsing step at the end of almost every cleaning operation. The frequent use of water makes it imperative that it contain minimum amounts of potentially harmful contaminants. The types of contaminants that may exist even in pure drinking water, but cannot be tolerated in water used for microelectronics include:

1. Dissolved inorganic salt such as sodium, and calcium salts. They are dissolved by the water as it flows through pipes, rocks, soils, etc.
2. Dissolved organic compounds from industrial waste or living matter.
3. Particulate matter such as small silica particles from rock, soil, and paper.
4. Microbiological life that sustains itself on other contaminants.

Water containing these types of impurities must be purified to reach the levels in Table 16-1.

TABLE 16-1: High Purity Semiconductor Manufacturing Water
Compared to Tap Water

Water Specification	Tap Water	High Purity Water
resistivity (megohm-cm)	.0002	15–18
electrolytes (parts per billion)	200,000	<25
particulate content (#/cm^3)	100,000	<150
living organisms (#/cm^3)	100–10,000	<10

The process of ion exchange or deionization was used almost exclusively in water purification prior to the last few years. Ion exchange is the removal of positive and negative ions using activated resins. A typical ion exchange water system is shown in Figure 16-1. Such a system contains the following elements:

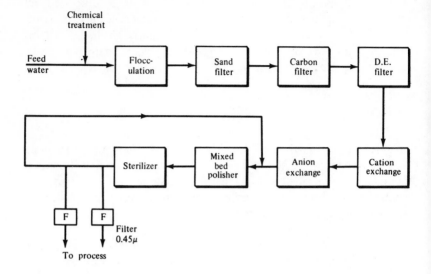

Figure 16-1 Typical ion exchange high purity water system.

1. *Chemical treatment* (often chlorination). Kills organisms present in the feed water.
2. *Sand filter*. Removes particles from the incoming water.
3. *Activated charcoal filter*. Removes any free chlorine and traces of organic matter.
4. *Diatomaceous earth filter*. Retains additional contaminants.
5. *Anion exchange*. Strongly ionized acids such as sulfuric, hydrochloric, and nitric are removed.
6. *Mixed bed polisher*. Contains both cation and anion resins and removes any ions missed by previous exchange filters.
7. *Sterilization*. Bacteria growth is controlled by one of several methods such as chlorination or ultraviolet light.
8. *Filter*. Any residual particles are removed prior to using the wafer.

Systems like the ones described above have been partially displaced by reverse osmosis or RO systems. Under pressure and in the presence of a selectively permeable membrane, water will flow through the membrane, while dissolved or suspended substances will not. A diagram of a typical reverse osmosis water system is shown in Figure 16-2.

The differences between this system and the ion exchange sys-

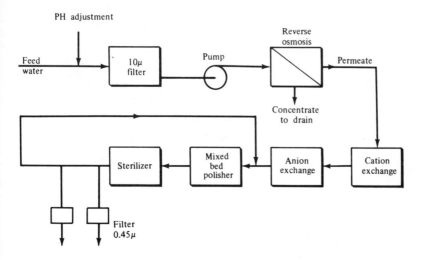

Figure 16-2   Typical reverse osmosis high purity water system.

tem are the pH adjustment, the filtering, and the use of reverse osmosis. All of the other steps are identical. Reverse osmosis is effective because it reduces the frequency with which the ion exchange resins must be regenerated.

Once water has been purified to the required level, it is necessary to distribute it throughout the facility without suffering losses in water quality. This requirement is usually met by running the water in a type of inert plastic to prevent any significant recontamination.

16-3   AIR

The three major parameters that must be controlled in the ambient air are temperature, humidity, and particle level. The temperature and humidity are determined by the type, amount, and set point of the air handling equipment, but the presence of particles must be controlled in another fashion. Particles prove to be more detrimental to the process at certain steps such as during cleaning and loading operations or during the photoresist process. Accordingly, if efforts are made to control the presence of unwanted airborne particles at the critical operations, the process will not suffer. This philosophy brought about the use of laminar flow hoods or stations at certain operations in device fabrication. The cross section of a vertical laminar flow hood is shown in Figure 16-3.

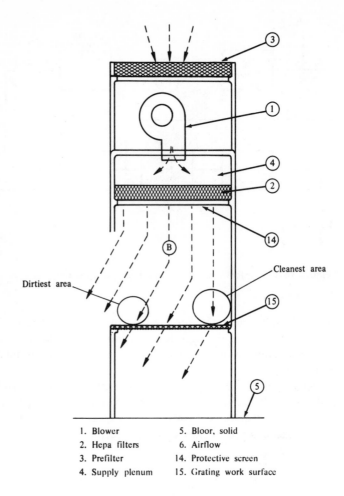

1. Blower   5. Bloor, solid
2. Hepa filters   6. Airflow
3. Prefilter   14. Protective screen
4. Supply plenum   15. Grating work surface

Figure 16-3   Vertical laminar flow hood.

A laminar flow uses the room as a reservoir, taking air from the room, filtering it, and blowing it into the work area in a parallel or laminar pattern. Laminar flow prevents the formation of turbulent regions that may accumulate high levels of contamination. The most important part of the laminar flow hood is the *H*igh *E*fficiency *P*articu-late *A*ir or HEPA filter. A HEPA filter is a fragile but effective filter capable of removing a minimum of 99.97 percent of all $.3\mu$ or larger particles. Use of tested HEPA filters in all laminar flow hoods is a necessity.

16-4 GASES

The gases needed in semiconductor fabrication are nitrogen, oxygen, hydrogen, HCl, ammonia, and occasionally some others. Precautions must be taken to insure that no contamination reaches the wafers through these gases. Copper tubing may be used for oxygen and nitrogen, though stainless steel may prevent the occurrence of certain problems. Stainless steel tubing should be used for all other gases, because of its high resistance to corrosion. Filters are an important item, and should be used near the point of use to prevent the passage of any unwanted particles.

16-5 PERSONNEL/CLEAN ROOM

The greatest sources of contamination in a wafer fabrication area are the people that perform the operations. Human beings are a continuous source of organic matter because of the constantly renewing nature of their bodies. It is possible to neutralize the presence of operations by instituting certain sets of procedures. The first step is to partition off the wafer fabrication area and control access to it. Smocks are often worn in a fabrication area, but they do little more than serve as a protective coating for the operator. Their impact on contamination control is minimal. The most effective method of controlling contamination from operators is to provide them with clothing that covers as much of their body as possible including an outer covering on the hands and the feet. These suits must be cleaned regularly to keep the level of accumulated contamination low.

REVIEW EXERCISES: CONTAMINATION CONTROL

1. What chemical, present in the human body, is a mobile contaminant in silicon dioxide?
2. State two methods of obtaining high purity water for semiconductor processing.
3. For wafers of unknown origin, should the acid clean or the solvent clean be first? Why?
4. Why are laminar flow hoods placed just over critical areas in a wafer fabrication area?

5. List and describe the common cleaning steps utilized to prepare wafers for fabrication into devices.

6. Sketch and label a typical ion exchange water purification process.

7. List the major differences between the ion exchange and reverse osmosis water purification systems.

8. How is purified water distributed to various points within a semiconductor processing facility?

9. What metal or metals are frequently utilized to transport the various gasses required during semiconductor processing?

10. Sketch and describe a typical wafer fabrication area as to contamination control.

17

Advanced Silicon Technology

Silicon technology emerged as the dominant force in solid state devices in the early 1960s and has continued to progress at a rapid rate. This progress has been the result of a combination of better understanding of the materials, the devices, and the processing steps involved in their fabrication. The net result of this tremendous progress in silicon technology has been an increase in the types of functions that can be performed with silicon devices, coupled with a decrease in the cost per function. Silicon technology seems likely to continue in the manner it has over the next few decades, with a steady progression of advances along a number of fronts. In this section we look at emerging technologies and attempt to assess their impact on the main stream of solid-state technology.

17-1 DOMINANT TRENDS IN TECHNOLOGY: SUBSTRATE SIZE AND DEVICE DENSITY

The two trends in silicon technology that have significantly decreased the cost per function of circuits have been the use of continually larger wafers, and the ability to manufacture devices with smaller geometries, and hence higher packing densities.

The introduction of larger silicon substrates every few years has served to continually push processing technology forward. To prevent a decrease in yield every time a larger substrate size was introduced, processing technology had to undergo improvements to maintain the same uniformity across these larger wafers. Economics is the reason for the use of ever-larger wafers—it costs only slightly more to process each larger wafer, but the total number of good die with the same yield percentage is proportional to the increase in area. The ability of crystal growers to produce ever-larger wafers seems to be a continuing phenomenon and one wonders about the size of the silicon wafers in the future. The use of ever-larger wafers requires that the thickness of the wafers also be increased to have wafers with the same resistance to breakage. These wafers must be sawed from crystals, a process in which two-thirds of the grown crystal is discarded. An alternative to silicon wafers grown in the conventional manner is available —using a crystal growth technology known as *E*dge-defined *F*ilm-fed *G*rowth (EFG). Using this technique, it may be possible to grow continuous ribbons of silicon of a preselected width and thickness. These ribbons of silicon could be cut into substrates of a required size, and processed with only a minimum of wafer preparation, resulting in a significant savings in the substrate cost.

Silicon ribbons produced using the EFG technique are already being tested for possible use as the substrates for photodiode fabrication. If this technique produces silicon of sufficient quality for photodiodes at an economically attractive level, these photodiodes will be used to convert the sun's radiation into electricity on a continually expanding basis. From that time onward, the relative impact of EFG silicon will be determined by a complex series of events. The use of round wafers will be a well-established process, and the ability of a new substrate fabrication process to impact a well-established process is impossible to foresee.

17-2   ALIGNMENT/EXPOSURE STEP

The alignment/exposure step discussed in the photomasking section is another area in which significant technological strides are being made. In conventional alignment systems, the wavelength of the light used to expose the photoresist sets a limitation on the minimum dimensions that can be transferred from a mask to a wafer because of the diffraction of the light. (Diffraction is the bending of light as it goes past an edge.) If any space is left between the mask and the wafer during exposure, diffraction will occur. The minimum dimensions that could be transferred if everything else were perfect is about the wavelength of the light used. However, practical problems such as wafer flatness, mask flatness, particles, etc., make such dimensions impossible in a production situation. The wavelength of the illumination from the mercury arc lamp is about 4000 Å, so a practical minimum for the line width using optical techniques is between $.5\mu$ and 1.0μ. A survey of contemporary technology shows that some devices are within a factor of 5 of these dimensions already.

The fundamental limitation is the wavelength of the light used to expose the photoresist, so exposure using a shorter wavelength offers a possible solution to this problem. Two possible alternative methods of exposing a "sensitized" resist layer would be to use either x-rays or electrons.

X-rays generated from a number of sources have low enough energy not to pose a long-term problem to human beings. The wavelengths of these "soft" x-rays range from 5–15 Å. This wavelength is also short enough not to impose any serious diffraction problems. With a source of x-rays and an x-ray sensitive resist, all that is needed is a mask and a method of aligning the mask to the wafer. A diagram of an x-ray exposure system is shown in Figure 17-1.

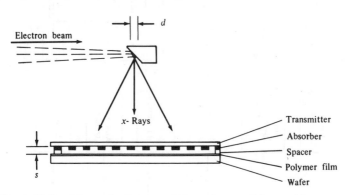

Figure 17-1   Schematic diagram of the soft x-ray lithographic system.

The generation of a mask for such a system, and the problem of aligning the mask to an image already present on the wafer are subjects which are presently being researched. The entire operation is complicated by the need to expose the wafers in a vacuum because of the limited distance the x-rays travel in air.

The use of electron-beam exposure offers many of the advantages of x-ray exposure, as well as a few others. Electrons are charged particles, but they can be viewed as having wave properties. The equivalent wavelength of electrons used for exposure systems is less than an angstrom. Electron beams can be generated using contemporary equipment, leaving the problem of finding an electron-sensitive resist, obtaining a mask for the exposure, and aligning this mask to an already

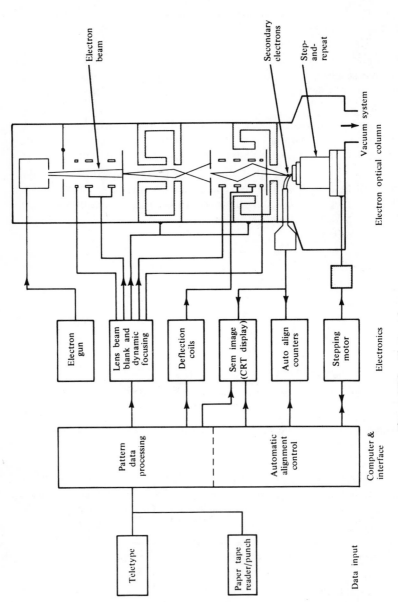

Figure 17-2 Electron-beam exposure system.

existing pattern on a wafer. A diagram of such an exposure system is shown in Figure 17-2.

Thus far, x-ray and electron-beam exposure systems look quite comparable, but there are two advantages to an electron-beam exposure system that have not been mentioned. First, the electron-beam microscope is an instrument used for high magnifications. Modifying the electron-beam system for alignment as well as magnification solves the mask-to-wafer alignment problem. Second, because an electron-beam is a stream of charged particles, it is possible to deflect this beam as well as to turn the beam on and off. This ability means that no mask is really necessary to expose a layer of resistive resist. The beam can be scanned and turned on and off to directly accomplish the exposure. The potential of exposing wafers by using a computer-controlled scanning electron-beam is an attractive possibility.

Both x-ray and electron-beam exposure systems offer potential for the future. But, both types of systems need considerable development before either of them will be used in large-scale production.

17-3   DEVELOPMENTS IN PROCESSING TECHNOLOGY

Processing technology has been marked by a series of steady improvements in all areas. One major development that has impacted both yields and the circuits that are being fabricated is the controlled oxidation of selected regions of silicon. This technology is exemplified in Fairchild's Isoplanar process. In this process, silicon nitride is deposited on the wafers of an epitaxial silicon wafer with a thin layer of silicon dioxide on it. The silicon nitride is masked and etched leaving the wafer as shown in Figure 17-3a. Next, the silicon epitaxial layer is etched about half-way through in regions not covered with silicon nitride as shown in Figure 17-3b. Finally, this wafer is placed in an oxidation furnace until the growing oxide layer is through the epitaxial layer. Since oxidation of a 1μ thick layer of silicon produces about a 2μ layer of silicon dioxide, the silicon dioxide region is level with the top surface of the wafer while separating regions of the epitaxial layer from each other as shown in Figure 17-3c.

The slow oxidation of silicon nitride compared to silicon is the key factor in this process step. The use of a region of silicon dioxide to separate active devices, instead of silicon of the opposite conductivity type, means that devices may be packed much more densely. The use of a similar technique in MOS circuits results in a corresponding savings in area, while making it easier to manufacture the devices.

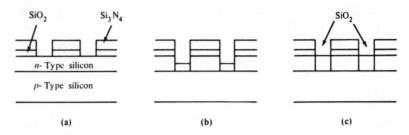

Figure 17-3   Local oxidation of silicon; **(a)** nitride removal; **(b)** silicon removal; **(c)** final.

17-4   DEVELOPMENTS IN DEVICE TECHNOLOGY

Two recent developments in device technology show the possibility of significantly impacting the direction of technology. The first of these developments was the discovery of the charge-coupled device or the CCD. In its simplest form, a CCD consists of a lightly doped silicon substrate with a thin layer of SiO_2 on it, and a series of metallized electrodes on top of the oxide layer as shown in Figure 17-4. The charge is stored in potential "wells" created by the voltage on the electrodes. To transfer the charge, a deeper potential well is created next to the storage site, causing the carriers to fall into the next well as shown in Figure 17-5. By varying the potential along a series of electrodes, the "packet" of charge can be made to move from one location to another on the semiconductor chip.

The relative simplicity of the CCD structure means that few fabrication steps are required in its manufacture. Fewer fabrication steps imply that large-area circuits can be economically made. Circuits that require large areas include digital memories, optical sensors, and signal processors. Application of CCD technology to these three areas is already underway, resulting in a series of successful products.

Integrated Injection Logic or I^2L is the second developing technology with great promise for the future. I^2L is a logical extension of bipolar design technology that greatly reduces the area needed to perform many digital functions. This technology was eventually standard bipolar processing, but uses inverted vertical *npn* transistors. As shown in Figure 17-6, the common emitter is the n^+ region beneath the entire structure. The use of the substrate as a common emitter uses a minimum of chip area. Separate collector regions are necessary for circuit application. A lateral *pnp* transistor is also part of the basic I^2L circuit, supplying current to the bases of the *npn* transistors as shown in Figure 17-6.

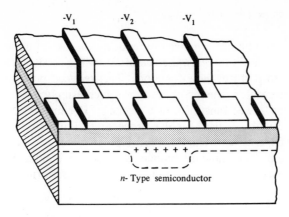

Figure 17-4 A CCD in the storage mode. $-V_2$ is greater than $-V_1$, forming a potential well that captures charge.

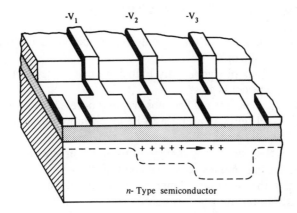

Figure 17-5 Charge is transferred to the next electrode when a still larger voltage $-V_3$ creates an even deeper potential well.

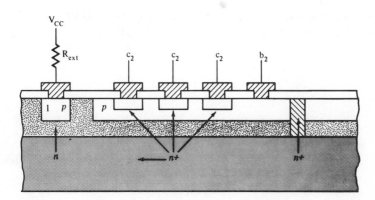

Figure 17-6 Cross section of a I^2L circuit.

In addition to significantly reducing the size of circuits required to perform digital functions, the use of I²L will also significantly reduce the power to perform these functions. This reduced power requirement will in turn result in the incorporation of increasingly larger numbers of circuits, and hence functions, in a package. Because I²L processing is compatible with standard bipolar processing, future circuits may use a mixture of I²L circuits and standard bipolar circuits to accomplish circuit function. The standard bipolar circuits can handle the input and output devices, while the denser I²L devices perform the necessary operations.

REVIEW EXERCISES: ADVANCED SILICON TECHNOLOGY

1. What is the name of the crystal growth method used to grow thin ribbons of silicon?
2. Two methods of significantly reducing device sizes involve the use of what types of beams?
3. Does silicon nitride oxidize more rapidly or less rapidly than silicon?
4. Why do I²L circuits promise to be much more dense than conventional bipolar circuits?
5. What is the principle advantage of using larger silicon wafers? What limits the maximum size of a silicon wafer?
6. What is the minimum dimension that could be transferred from mask to wafer using yellow light?
7. What is the minimum practical dimension that can be transferred from mask to wafer using today's semiconductor technology?
8. What are two major advantages of the electron-beam exposure system?
9. Sketch and explain the operation of a typical CCD structure.
10. What is a major advantage of CCD technology?

18

Nonsilicon Technology

The processing technology that has been discussed in the previous lessons can be applied to nonsilicon semiconductors and to nonsemiconductor materials. This lesson considers the application of the processing technology that has been discussed to nonsilicon devices.

18-1 LIGHT-EMITTING DIODES (LED'S)

Light-emitting diodes are fabricated in III-V semiconductor compounds such as gallium arsenide and gallium arsenide-phosphide. These semiconductors are called III-V semiconductors because one of the constituents of the compound is from group III of the Periodic Table while the other constituent is from group V. The electron configuration of certain of these compounds makes the emission of light possible when current flows through a diode that has been properly fabricated. The technology used to fabricate diodes in these semiconductors closely parallels standard silicon technology. The process steps are as follows:

1. Starting with the proper substrate, grow a carefully tailored layer of epitaxy.
2. Cover the front surface of the substrate with a protective layer of silicon dioxide using a low-temperature chemical vapor deposition technique.

3. Transfer an image to the front surface of the wafer by using a photolithographic process:
   a. Apply a layer of photoresist.
   b. Transfer a pattern to the photoresist using a mask.
   c. Using the photoresist as a mask, remove the deposited silicon dioxide in selected regions by etching.
4. Perform a high-temperature diffusion to form the diode.
5. Form ohmic contacts to the diode.
6. Separate the diodes on a substrate into discrete diodes or arrays.
7. Test and package the light emitting diodes.

18-2 OPTICAL INTEGRATED CIRCUITS

One long-range use of light-emitting diodes may be in optical integrated circuits. At the same time that light-emitting regions are fabricated in a substrate, it is possible to fabricate light-sensitive regions. Recent work indicates that it may be possible to link a light-emitting and a light-sensitive region with an intermediate region that serves as a light "wave-guide." With the successful fabrication of light generation, light transmission, and light reception regions on the same substrate, the fabrication of an optical integrated circuit will be possible. It will then remain for engineers and scientists to prove the usefulness of such circuits.

18-3 LIQUID CRYSTAL DISPLAYS

A rival to LED's in many applications, and particularly in the watch display market, is the liquid crystal display or LCD. The name of these devices describes their composition—LCD's have a liquid crystal material sandwiched between two glass plates separated by about 12μ. The glass plates are sealed around their perimeter. The image appears between two layers of conductive material, one etched on the surface of each glass plate.

In operation, a voltage is applied between the two layers of conductive material which changes the molecular orientation of the liquid crystal material. The light passing through the display is thus scattered or reflected in a nonequal manner where the voltage has been applied. LCD's emit no light, but can use either reflected or transmitted light to form the display. LCD's have the large advantage of consuming considerably less current than LED's.

18-4   QUARTZ CRYSTAL OSCILLATORS

Silicon dioxide in its crystalline form or quartz is a piezoelectric material. When it is placed in an electric circuit with a voltage across it, it vibrates producing an output whose frequency is a function of the physical parameters of the crystal. Since quartz is silicon dioxide, it is not surprising to find the type of technology discussed earlier used in the manufacture of timing devices from quartz. The specific steps used in their manufacture are centered around the transfer of an image to both sides of a quartz wafer and a subsequent etching step to remove the silicon dioxide in unnecessary areas. The ability to produce large numbers of quartz timers simultaneously is the main reason for the use of microelectronics technology.

18-5   MAGNETIC BUBBLE OR MAGNETIC DOMAIN DEVICES

Within the past ten years, a phenomenon in which magnetic "bubbles" are formed and can be made to move and interact in predetermined ways has been discovered. The bubbles are actually cylindrical magnetic domains whose polarization is opposite to that of the thin magnetic film in which they are embedded. The bubbles are stable over a wide range of conditions, and can be moved from one point to another at high velocity. The technology used to form the control regions on the surfaces of the thin magnetic film are similar to those used in the fabrication of integrated circuits. The present evidence indicates that memories made using magnetic bubbles may be a rival to the use of a standard semiconductor memories in many applications.

18-6   HYBRID TECHNOLOGY

Hybrid technology is the technology of placing devices and circuits on a common substrate and interconnecting them to perform useful electrical functions. Hybrid technology can be divided into two fabrication technologies: thin film and thick film.

Thin-film hybrid circuits are manufactured by first depositing a thin layer of metal on a substrate using vacuum deposition techniques. Next, the substrates are coated with photoresist, baked, exposed, and the metal layer is selectively removed by etching to form the desired pattern. Active and passive devices such as diodes, resistors, transistors, and capacitors are attached to the substrate to complete the circuit.

Thick-film structures are prepared by screening and firing or by pyrolitic deposition. They generally contain only conductors, resistors, and capacitors, with the other components added as discrete entities. All are put down on a substrate which is generally composed of some sort of alumina. A thick film is a conductive, resistive, or insulating film thicker than 10 mils that is produced by firing a paste deposited on a substrate. The paste is deposited on the substrate by the stencil screen process. Different paste compositions are used for each component in the circuit. After the pattern has been screened onto a ceramic substrate and dried, it is fired in a furnace, where the composition of the paste gives rise to the final characteristics of the layers.

Although thick-film technology is usually less costly than thin-film, it requires a larger substrate to accommodate the same circuit complexity. Generally speaking, a thick-film circuit will be limited in resistor tolerances to no better than ±1% and resistances to less than 5 megaohms. Thick-film techniques are generally used to build circuits operating below 1 Gigahertz, and which do not require the tolerances and line precision obtainable with thin-film techniques. Thin-film fabrication, on the other hand, is ideal for high-frequency microwave applications, and those requiring highly precise line widths and circuit elements.

REVIEW EXERCISES: NONSILICON TECHNOLOGY

1. Does a LED or a LCD display require more power?
2. What effect is responsible for the use of quartz as a timer?
3. What is the basic difference between thin-film and thick-film hybrid circuits?
4. List and describe the processing steps required to produce light-emitting diodes.

Scientific Notation

AI-1 MATHEMATICS

In dealing with physically measureable amounts of materials, very large and very small numbers are usually encountered. The writing of these numbers usually means using many zeros. To avoid the continuous writing of zero, mathematicians devised a shorthand method usually referred to as scientific notation. The idea behind scientific notation is built around the number ten, and can be developed as follows. The number 10 can also be written as 10 to the first power or 10^1. The number 1 is the power to which 10 is raised. (The power to which a number is raised is the number of times that number is multiplied by itself.) If we multiply 10 by itself, we get $10 \times 10 = 100$. But this is two 10's multiplied together, so $10 \times 10 = 10^2$ (10 to the second power) $= 100$. In a similar fashion, $10 \times 10 \times 10 = 10^3 = 1,000$. A power of 10 greater than zero tells us the number of zeros to the right of the 1 in the number we are dealing with. Listed below are some common multiples of ten with powers greater than zero.

$$10 = 10^1 = \text{ten to the first power}$$
$$100 = 10^2 = \text{ten to the second power}$$
$$1,000 = 10^3 = \text{ten to the third power}$$
$$10,000 = 10^4 = \text{ten to the fourth power}$$
$$100,000 = 10^5 = \text{ten to the fifth power}$$
$$1,000,000 = 10^6 = \text{ten to the sixth power}$$

Numbers smaller than one can also be represented using scientific notation. The number $\frac{1}{10} = .1$ is written as 10^{-1} (ten to the minus one power). Similarly, the number $\frac{1}{100} = \frac{1}{10^2} = .01$ is written as 10^{-2} (ten to the minus two power). A power of ten less than zero tells the number of places the decimal point is to the left of the 1 in the number we are dealing with. More common multiples of ten with powers less than zero are listed here.

$$.000001 = 10^{-6} = \text{ten to the minus sixth power}$$
$$.00001 = 10^{-5} = \text{ten to the minus fifth power}$$
$$.0001 = 10^{-4} = \text{ten to the minus fourth power}$$
$$.001 = 10^{-3} = \text{ten to the minus third power}$$
$$.01 = 10^{-2} = \text{ten to the minus second power}$$
$$.1 = 10^{-1} = \text{ten to the minus first power}$$

In most cases, it is convenient to rewrite large or small numbers as numbers between one and ten times ten to some power. To express a number greater than ten in this form, move the decimal one place to the left and count the number of places from the original decimal point. The number of places counted will give the correct positive power of 10.

Examples:

$$942 = 9.42 \times 10^2$$
$$420,610 = 4.2061 \times 10^5$$
$$31 = 3.1 \times 10^1$$

To express a number less than 1 in this form, move the decimal point to the right until it is part of the first nonzero number. Count the number of places the decimal point has been moved from its original position. The number of places counted is the negative power of ten.

Examples:

$$.00882 = 8.82 \times 10^{-3}$$
$$.0000031 = 3.1 \times 10^{-6}$$
$$.064 = 6.4 \times 10^{-2}$$

The only power of ten we have not discussed so far is 10^0. The value of 10^0 is 1. This statement makes sense when considering the values of 10^{-1} and 10^1.

$$10^{-1} = \quad .1$$
$$10^0 \ = \ 1$$
$$10^1 \ = 10.$$

AI-2 ADDITION AND SUBTRACTION

Before numbers written using scientific notation can be added or subtracted, all numbers must be written using the same power of ten. Once the power of ten of the numbers is the same, addition and subtraction is performed, and the answer is the resulting number to the common power of 10.

Example 1:

$$4.13 \times 10^3$$
$$+9.64 \times 10^5$$

First, rewrite 9.64×10^5 as 964×10^3. Then rewrite the problem.

$$4.13 \times 10^3$$
$$+964.00 \times 10^3$$
$$\overline{\ 968.13 \times 10^3} = 968{,}130$$

This solution can be checked, since

$$4.13 \times 10^3 = 4130 \text{ and } 9.64 \times 10^5 = 964000$$

This sum is just

```
  964000
+   4130
  ———————
  968130
```

Example 2:

$$4.93 \times 10^7$$
$$-9.4 \ \times 10^5$$

At first glance, it appears that the subtraction will result in a number less than zero. But, the first number is rewritten as:

$$4.93 \times 10^7 = 493 \times 10^5$$

The subtraction is done, keeping the common power.

$$\begin{array}{r} 493 \times 10^5 \\ - 9.4 \times 10^5 \\ \hline 483.6 \times 10^5 \end{array}$$

AI-3 MULTIPLICATION

Multiplication of numbers written in scientific notation is accomplished using the following rules:

> **1.** Multiply the number to the left of the power of 10 to obtain the numerical part of the product.
> **2.** Add the powers of ten together to get the resultant power of 10.

Example 1:

$4.3 \times 7.6 = 32.68$

$$\left.\begin{array}{l} 4.3 \times 7.6 = 3268 \\ 10^4 \times 10^2 = 10^6 \end{array}\right\} = 32.68 \times 10^6 = 3.268 \times 10^7$$

Example 2:

$1.2 \times 9.1 = 10.92$

$$\left.\begin{array}{l} 1.2 \times 9.1 = 1092 \\ 10^{-4} \times 10^6 = 10^2 \end{array}\right\} = 10.92 \times 10^2 = 1.092 \times 10^3$$

AI-4 DIVISION

The rules for division of numbers written using scientific rotation are given below:

> **1.** Divide the numbers to the left of the power of 10 to obtain the numerical portion of the answer.
> **2.** Subtract the power of 10 of the denominator from the power of 10 of the numerator to obtain the power of 10 of the answer.

Example 1:

$$\frac{9.81 \times 10^7}{4.19 \times 10^3}$$

$$\left.\begin{array}{l} \dfrac{9.81}{4.14} = 2.34 \\[3mm] \dfrac{10^7}{10^3} = 10^{7-3} = 10^4 \end{array}\right\} = 2.34 \times 10^4$$

Example 2:

$$\frac{3.1 \times 10^{-3}}{5.6 \times 10^3}$$

$$\left.\begin{array}{l} \dfrac{3.1}{5.6} = .554 \\[3mm] \dfrac{10^{-3}}{10^3} = 10^{-3-(3)} = 10^{-6} \end{array}\right\} = .554 \times 10^{-6} = 5.54 \times 10^{-7}$$

Use of Graphs

Throughout the text lessons, much of the information presented is in the form of graphs. Graphs are a way of presenting much information in a small space, without getting into the complicated mathematics often involved. A typical graph is shown in Figure AII-1.

This graph has time in minutes along the horizontal axis and distance in miles along the vertical axis. If a car is traveling at 60 miles per hour (or 1 mile per minute), the line represents the distance traveled for any elapsed time. To determine the distance covered in 30 minutes, find 30 minutes on the horizontal axis, and follow a path straight upward until it intersects the line. Then follow a path straight across to the vertical axis. The point of intersection should be at 30 miles along this axis. In a similar fashion, the distance traveled for any time up to 100 minutes can be found using this graph. The information provided by Figure AII-1 can be easily obtained by multiplying the time elapsed by one mile per minute to obtain the total distance traveled. But if a constant speed is not used, a graph such as that shown in Figure AII-2 would result. At the end of 100 minutes, 100 miles have been covered, but at times between 0 and 100 minutes, a set relationship between the time elapsed and the distance traveled is not easily obtained. This graph provides a method of determining the relationship between the time elapsed and the distance traveled without requiring the use of complicated mathematics.

There are three types of graphs that we will be concerned with in these lessons. These three types of graphs have different types of coordinate scales along each axis. The first type of graph has the variables

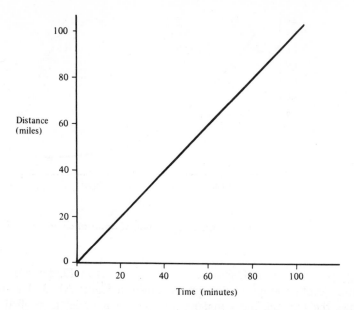

Figure AII-1   Graph of the distance traveled by a car vs. time for a constant velocity.

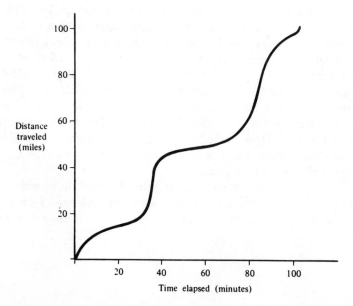

Figure AII-2   Graph of the distance traveled by a car vs. time for a varying velocity.

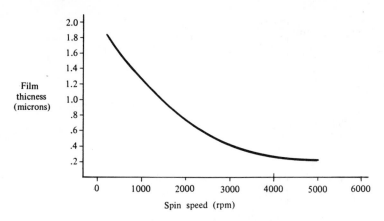

Figure AII-3   Film thickness as a function of application spin speed.

displayed in a linear fashion along each axis as in Figure AII-1 and AII-2. Another graph of this type is shown in Figure AII-3. This figure shows the film thickness that results when a material is applied to a wafer using a spinning technique. Both axes have linearly increasing variables, usually starting with zero.

The second type of graph has the variable along one axis displayed using a logarithmic scale as shown in Figure AII-4. The horizontal scale is linear, but the vertical scale increases by a factor of 10 between every major line. Use of a logarithmic scale allows information with a large numerical range to be displayed. An expanded portion of the logarithmic side is shown in Figure AII-5. In Figure AII-4, the numbers between 10^{17} and 10^{18} are numbered 2,3, etc. The line at 2 corresponds to 2×10^{17}. The next line corresponds to 3×10^{17}, and so on until we reach 9×10^{17}. The line above 9×10^{17} is 10×10^{17} which equals 1×10^{18}. It is marked on the graph. In a similar fashion, the line marked 2 above the 1×10^{18} line is 2×10^{18}. This type of graph is often called a semi-log graph.

Information relating two variables that both have wide ranges can be displayed using a logarithmic scale along each axis. Such a graph is called a log-log graph, and an example of one is shown in Figure 8-1.

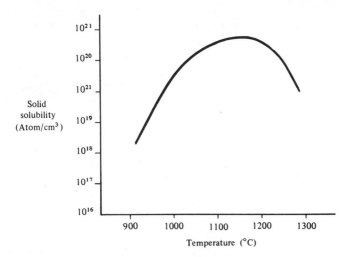

Figure AII-4   Solid solubility of elements in silicon as a function of temperature

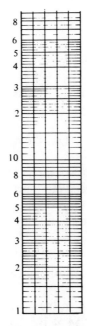

Figure AII-5   Extended section of the logarithmic scale.

Units

The semiconductor industry is presently undergoing a transition from the use of English units (inches, feet, mils) to the use of metric units (millimeters, angstroms, centimeters). The set of units encountered is a function of the area of the company a person works in. In many semiconductor fabrication areas, the surface measurements such as die size and device dimensions are measured in mils, while the junction depths and epitaxial layer thicknesses are measured in angstroms or microns. The ability to rapidly convert from one system of units to another is a necessity. Table AIII-1 is designed to assist in the conversion from one set of units to another. To accomplish conversion, find the first set of units on the left side and the second set of units along the top. The number at the intersection of the row and column is the conversion factor.

Example 1: Convert 2.3 mils to microns.

Solution: Following the rule from above, the number at the intersection of the "mil" row and the "micron" column is 25.4

$$2.3 \text{ mils} \times 25.4 \ \frac{\text{microns}}{\text{mil}} = 58.4 \text{ microns}$$

Example 2: Convert 32 microinches to angstroms

Solution: The number at the intersection of the microinch row and the angstrom column is 2.54×10^2. So

$$(32) (2.54 \times 10^2) = 8128 \text{ Å} = .8128 \text{ microns.}$$

TABLE AIII-1: Length Units Used in Semiconductor Technology

MULTIPLY BY (FROM) → TO GET ↓	INCH (")	MIL	MICROINCH	CENTIMETER (cm)	MILLIMETER (mm)	MICRON (μ)	ANGSTROM (Å)
INCH (")	1	10^3	10^6	2.54	25.4	2.54×10^4	2.54×10^8
MIL	10^{-3}	1	10^3	2.54×10^{-3}	2.54×10^{-2}	25.4	2.54×10^5
MICROINCH	10^{-6}	10^{-3}	1	2.54×10^{-6}	2.54×10^{-5}	2.54×10^{-2}	2.54×10^2
CENTIMETER (cm)	0.3937	3.937×10^2	3.937×10^5	1	10	10^4	10^8
MILLIMETER (mm)	3.937×10^{-2}	39.37	3.937×10^4	0.1	1	10^3	10^7
MICRON (μ)	3.937×10^{-5}	3.937×10^{-2}	39.37	10^{-4}	10^{-3}	1	10^4
ANGSTROM (Å)	3.937×10^{-9}	3.937×10^{-6}	3.937×10^{-3}	10^{-8}	10^{-7}	10^{-4}	1

Glossary

A-30: A commercial formulation of chemicals used to remove photoresist from wafers following an etching step. A-30 may be used with many metals.

Acetic acid (CH_3COOH): A weak acid often used in conjunction with a strong acid in cleaning and etching solutions.

Alloy: In semiconductor processing, the alloy step causes the interdiffusion of the semiconductor and the material on top of it, forming an ohmic contact between them.

Aluminum: The metal most often used in semiconductor technology to form the interconnects between devices on a chip. It is usually deposited by evaporation.

Ammonia (NH_3): A gas often used to react with silicon to form silicon nitride.

Ammonium fluoride (NH_4F): A chemical often used with hydrofluoric acid as a buffering agent to form etches for silicon dioxide.

Angle lap: A method for magnifying the depth of a junction by cutting (lapping) through it at an angle away from the perpendicular.

Angstrom: A unit of length. An angstrom is one ten-thousandth of a micron (10^{-4} microns).

Anneal: A high-temperature processing step (usually the last one) designed to minimize surface effects in devices by relieving stress or annealing the wafers.

Antimony (Sb): A Group V element that is an n-type dopant in silicon. It is often used as the dopant for the buried layer.

Arsenic (As): An n-type dopant often used for the buried layer predeposition.

Arsine (AsH_3): A gas that is often used as a source of arsenic for doping silicon.

Aqua regia: A mixture of nitric and hydrochloric acids often used to clean silicon wafers.

B: See boron.

Base: 1. The control portion of an *npn* or a *pnp* transistor. **2.** The *p*-type diffusion done using boron that forms the base of *npn* transistors, the emitter and collector of lateral *pnp* transistors, and resistors.

BCl₃: See boron trichloride.

Boat: 1. Pieces of quartz joined together to form a supporting structure for wafers during high-temperature processing steps. **2.** A Teflon or plastic assemblage used to hold wafers during wet processing steps.

Boat puller: A mechanical arrangement to push a boat loaded with wafers into a furnace and/or withdraw it at a fixed speed.

Bonding pad: The relatively large, circular, rectangular, or square areas of metallization that are probed or attached to when access to devices or circuits is desired.

Boron: The *p*-type dopant commonly used for the isolation and base diffusion in standard bipolar integrated circuit processing.

Boron trichloride (BCl₃): A gas that is often used as a source of boron for doping silicon.

Buffer: An additive that prevents the rapid change of the chemical activity of an acid or a base solution by keeping the number of ions capable of reacting essentially constant even as the solution is used.

Bump technology: A method of forming raised regions of metal over bonding pads to allow the simultaneous bonding of the "bumps" to a substrate or a package.

Buried layer: The $n+$ diffusion in the *p*-type substrate done just prior to growing the epitaxial layer. The buried layer provides a low-resistance path for current flowing in a device. Common buried-layer dopants are antimony and arsenic.

Capacitance: A measure of the amount of charge (DC) which a device can store in the dielectric between two conductors when a given voltage is applied. Capacitance is measured in farads.

Channel: A thin region of a semiconductor that supports conduction. A channel may occur at a surface or in the bulk. They may indicate contamination problems or incomplete isolation, if not wanted, but are essential for the operation of MOSFET's and SIGFET's.

Chip: One of the individual circuits on a wafer.

Chrome (Cr): A metal often used to fabricate masks. Chrome does not wear out as fast as emulsion, so chrome masks last longer.

Contact: The regions of exposed silicon that are covered during the metallization process to provide electrical access to the devices.

Contamination: A general term used to describe unwanted material that adversely affects the physical or electrical characteristics of a semiconductor wafer.

Cross section: A magnified display of the structure of a device. Diffused junctions, metallization, and oxide layers are often shown in this manner.

Current: A measure of the number of charged particles passing a given point per unit time.

Curve tracer: A piece of electrical test equipment that displays the characteristics of a device visually on a screen.

Develop: A photoresist process, this step removes the photoresist from areas not defined by the mask and exposure step.

Diborane (B_2H_6): A gas that is often used as a source of boron for doping silicon.

Die: See chip.

Dielectric: A material that conducts no current when it has a voltage across it. Two dielectrics encountered in semiconductor processing are silicon dioxide and silicon nitride.

Diffusion: A process used in the production of semiconductors which introduces minute amounts of impurities into a substrate material such as silicon or germanium and permits the impurity to spread into the substrate. The process is very dependent on temperature and time.

Diode: A two-terminal device that allows current to flow in one direction but not in the other. A diode is present at the intersection of a p-type and an n-type region of a semiconductor.

Dopant: An element that alters the conductivity of a semiconductor by contributing either a hole or an electron to the conduction process. For silicon, the dopants are found in Group III and Group V.

Dry oxide: Thermal silicon dioxide grown using oxygen.

E–beam: See Electron beam.

Electron: A charged particle revolving around the nucleus of an atom. It can form bonds with other atoms or be lost, making the atom an ion.

Electron beam: A type of evaporation that uses the energy of a focused electron beam to provide the required heat.

Emitter: 1. The region of a transistor that serves as the source or input end for carriers. **2.** The n-type diffusion usually done using phosphorous that forms the emitter of npn transistors, the base contact of pnp transistors, the $n+$ contact of npn transistors, and low-value resistors.

Emulsion: The opaque portion of a mask made using light-sensitive silver compounds.

Epi: See epitaxial.

Epitaxial: Greek for "arranged upon." The growth of a single crystal semiconductor film upon a single crystal substrate. The n-type layer of silicon deposited on the substrate and buried layer is epitaxial silicon.

Etch: A process for removing material in a specified area through a chemical reaction.

Evaporation: A process step that uses heat to evaporate a material from a source and deposit it on wafers. Both electron-beam and filament evaporation are common in semiconductor processing.

Filament: A coiled piece of wire that is loaded with a material to be evaporated and heated by passing current through it.

Four-point probe: A piece of electrical equipment used to determine the sheet resistivity of a predeposition or a diffusion.

Furnace: A piece of equipment containing a resistance-heated element and a temperature controller. It is used to maintain a region of constant temperature with a controlled atmosphere for the processing of semiconductor devices.

Getter: A process by which unwanted impurities are segregated out of a material such as silicon or silicon dioxide.

H₂: *See* hydrogen.

HCl: *See* hydrochloric acid.

HF: *See* hydrofluoric acid.

HNO₃: *See* nitric acid.

Hole: The concept used to describe the movement of the "absence of an electron" through the crystal structure of a semiconductor.

Hydrochloric acid (HCl): A strong acid often used to clean silicon.

Hydrofluoric acid (HF): A strong acid used to etch silicon dioxide. It is often diluted or buffered before it is used.

Hydrogen (H_2): A gas used in semiconductor processing primarily as a carrier gas for high-temperature reaction steps like epitaxial silicon growth.

Hydrogen peroxide (H_2O_2): A chemical that is a strong oxidizing agent. It is often used with sulfuric acid to remove photoresist.

Ion: An atom that has either gained or lost electrons, making it a charged particle (either positive or negative).

Iron oxide (Fe_2O_3): A material used in making long-lasting masks. It has the added advantage of allowing some visible light through, resulting in a see-through mask.

Isolation mask: The second mask in standard bipolar integrated circuit fabrication. Boron is diffused into silicon in regions etched during the isolation photoresist process and electrically separates or isolates regions of silicon.

Isopropyl alcohol: A solvent often used in semiconductor processing for final rinsing and drying.

J-100: A commercial formulation of chemicals used to remove photoresist from wafers following an etching step. J-100 may be used with many metals.

Junction: The place at which the conductivity type of a material changes from *p*-type to *n*-type or vice versa.

Kit part: A device or a group of devices made electrically accessible by a separate metal mask stepped into the standard array of circuit metallization.

Leaky: A much-used term implying the presence of an unwanted current when a voltage is applied between two points.

Mask: A glass plate covered with an array of patterns used in the photomasking process. Each pattern consists of opaque or clear areas that respectively prevent or allow light through. The masks are aligned with existing patterns on silicon wafers and used to expose photoresist prior to etching either silicon dioxide or a metal. Masks may be emulsion, chrome, iron oxide, silicon, or a number of other materials.

Metallization: The layer of high-conductivity material (a metal) used to interconnect devices on a chip. Aluminum is most often used with silicon.

Methanol: *See* methyl alcohol.

Methyl alcohol: A solvent often used in semiconductor processing for removing other solvents, as a wetting agent, or as a final rinse.

Micron: A unit of length. 1 micron (μ) is one-millionth of a meter (10^{-6} meters).

Monolithic: Refers to the single silicon substrate in which an integrated circuit is constructed.

MOSFET (Metal Oxide Silicon Field-Effect Transistor): A device that works by inducing a conductive channel in silicon using a metal gate over a layer of oxide.

N$_2$: *See* nitrogen.

Negative resist: Photoresist that remains in areas that were not protected from exposure by the opaque regions of a mask while being removed in regions that were protected by the develop cycle. A negative image of the mask remains following the develop process. Waycoat and microneg are two common negative resists.

Nitric acid (HNO$_3$): A strong acid often used to clean silicon wafers or etch metals.

Nitride: *See* silicon nitride.

Nitrogen (N$_2$): A gas that seldom reacts with other materials. It is often used as a carrier gas for chemicals in semiconductor processing.

npn **transistor:** A transistor with an emitter and collector of *n*-type silicon and a base of *p*-type silicon.

n-**type:** A dopant belonging to the Group V elements. In silicon, the dopants fifth outer electron is free to conduct current.

O$_2$: *See* oxygen.

Ohm (Ω): The unit used to express resistance. One ohm is the resistance against which one volt will cause a current of one amp to flow.

Ohmic: A term used to denote a linear relationship between the voltage across a region and the current through it. An ohmic contact has this linear relationship, but hopefully the resistance is low.

1,1,1-Trichloroethane: A solvent which replaces regular trichloroethylene.

Operational amplifier (abbreviated op amp): The basic building block of linear circuits. The 709 and the 741 are operational amplifiers produced in large quantities for use in equipment like analog computers.

Oxide: *See* silicon dioxide.

Oxygen (O$_2$): A gas used in semiconductors to oxidize silicon, to form vapor-deposited oxide, and for other processing steps.

Passivation: A layer of a material put over an integrated circuit to stabilize the surface of its devices. Silicon dioxide or silicon nitride are often used for passivation.

Phosphine (PH$_3$): A gas that is often used as a source of phosphorus for doping silicon.

Phosphorus: The *n*-type dopant commonly used for the sinker and emitter diffusions in standard bipolar integrated circuit technology.

Phosphorus oxychloride (POCl$_3$): A liquid that is often used as a source of phosphorus for doping silicon.

Photoresist: The light-sensitive film spun onto wafers and "exposed" using high-intensity light through a mask. The "exposed" photoresist can be dissolved off of the wafer using developers leaving a pattern of photoresist which allows etching to take place in some areas while preventing it in others.

Plating: The electrochemical process used to deposit a metal on a desired object by placing the object at one electrical polarity and passing a current through a chemical solution to another electrode. The metal is plated from either the solution or the other electrode.

pnp **resistor:** A transistor with an emitter and collector of *p*-type silicon and a base of *n*-type silicon.

POCl$_3$: *See* phosphorus oxychloride.

Positive resist: Photoresist that is removed in areas that were not protected from exposure by the opaque regions of a mask while remaining in regions that were protected by the develop cycle. A positive image of the mask remains following the develop process. AZ-1350 is a common positive resist.

Poly: *See* polycrystalline silicon.

Polycrystalline silicon: Silicon composed of many (poly) crystals. Raw silicon comes in ingots of poly prior to crystal growth. Poly may be deposited epitaxially (either accidentally or on purpose) by depositing it too fast, at too low a temperature, or by depositing on a layer of silicon dioxide.

Predeposition (often called predep): The process step during which a controlled amount of a dopant is introduced into the crystal structure of a semiconductor.

p-type: A dopant belonging to the Group III-A elements. In silicon, the absence of a fourth outer electron manifests itself as conduction by a positively charged particle called a hole.

PVX (a shortened name for phosphorus-doped vapor-deposited oxide): a chemically deposited layer of phosphorous-rich silicon dioxide. PVX can be used for scratch protection, but is often used with a layer of vapox.

Quartz: Another name for silicon dioxide. Because of its temperature-resistant properties, quartz is used in many processing steps in integrated circuit fabrication.

Radio frequency: The energy medium used to heat the susceptor in most epitaxial reactors. Radio frequency means that the energy is transferred at a frequency near the normal radio transmitting band.

Reactor: A piece of equipment used for the deposition of a layer of material used in semiconductor processing. Common types of reactors are epitaxial reactors, vapor reactors, and nitride reactors.

Resistance: A measure of the difficulty in moving electrical current through a material when voltage is applied. Resistance is designated by the symbol R, and is measured in ohms.

RF: *See* radio frequency.

Sheet resistivity: A measurement with dimensions of ohms per square that tells the number of n-type or p-type donor atoms in a semiconductor.

SIGFET (Silicon Gate Field Effect Transistor): A device similar to a MOSFET but with a gate of doped polycrystalline silicon instead of metal.

Silane (SiH_4): A gas that readily decomposes into silicon and hydrogen. It is often used to deposit epitaxial silicon, and reacts with ammonia to form silicon nitride or oxygen to form silicon dioxide.

Silicon (Si): The Group IV element used for fabricating diodes, transistors, and integrated circuits.

Silicon dioxide (SiO_2): A passivating layer that can be thermally grown or deposited on silicon wafers. Thermal silicon dioxide is commonly grown using either oxygen (O_2) or water vapor (H_2O) at temperatures above 900°C.

Silicon nitride (Si_3N_4): A passivating layer chemically deposited on wafers at temperatures between

600°C and 900°C. It protects devices against contamination once it is applied.

Silicon tetrachloride ($SiCl_4$): A gas that reacts with hydrogen producing silicon and hydrogen chloride gas. It is often used to deposit epitaxial silicon.

Sinker: An n^+ diffusion from the surface of a device to the buried layer. The sinker provides a low-resistance path from the collector contact to the buried layer. Phosphorous is the dopant commonly used for sinkers.

Si_3N_4: *See* silicon nitride.

Slug: *See* buried layer.

Sputtering: A method of depositing a film of material on a desired object. A target of the desired material is bombarded with RF-excited ions which knock atoms from the target and deposit them on the object to be coated.

Steam oxide: Thermal silicon dioxide grown by bubbling a gas (usually oxygen or nitrogen) through water at 95°–98°C.

Subcollector: *See* buried layer.

Sulfuric acid (H_2SO_4): A strong acid often used to clean silicon wafers and to remove photoresist.

Susceptor: The flat slab of material (usually graphite) on which wafers are heated during high-temperature deposition processes like epitaxial growth or nitride deposition.

TCE: *See* trichloroethylene.

Thermal oxide: A layer of silicon dioxide grown in a furnace.

Thermocouple: A device to measure the temperature in a furnace or a reactor. It is made by welding two wires together at a point. Heat generates a voltage between the two materials that is proportional to the temperature.

Transistor: A three-terminal electrical device fabricated in silicon having three distinct regions:

a. emitter—the carriers originate here.

b. base—the control region

c. collector—carriers leave the transistors here

Transistors may be either *pnp* or *npn* devices.

Trichloroethylene: A solvent often used in semiconductor processing to remove grease or wax from wafers, boats, glassware, or other articles. Its use has been discontinued in many areas because of environmental considerations.

Tube: 1. See furnace. **2.** A cylindrical piece of quartz with fittings on one or both ends. It is placed in a furnace to provide a contamination-free and controlled atmosphere.

Vapox (a shortened name for vapor deposited oxide): A chemically deposited layer of silicon dioxide. It is usually deposited at temperatures between 350°C and 500°C, and is often used for scratch protection.

Voltage: The force applied between two points to try to cause charged particles (and hence current) to flow.

Wafer: A usually round, thin slice of a semiconductor material. Often used when referring to a wafer of silicon.

Wafer sort: The step at which integrated circuits are tested to see whether or not they work. Probes contact the pads of the circuit and they are measured by putting in an electrical signal and seeing if the correct one comes out.

Wet oxide: Thermal silicon dioxide grown by bubbling a gas (usually oxygen or nitrogen) through water at some temperature between 0°C and 100°C.

Xylene: A solvent often used in semiconductor processing to remove unexposed photoresist.

APPENDIX V

Solutions

1. Since only phosphorus atoms have been added and they are donors,

$N_D = 10^{15}/\text{cm}^3$ and $N_A = 0$

All of the donors ionize, with each donor producing one conduction electron. Therefore

$n = 10^{15}/\text{cm}^3$

We can solve for p, since $n = p = n_i^2$ and $n_i^2 = 2 \times 10^{20}/\text{cm}^6$

$$p = \frac{n_i^2}{n} = \frac{2 \times 10^{20}}{10^{15}}/\text{cm}^3 = 2 \times 10^5/\text{cm}^3$$

Figure 1-7 can be used to determine the resistivity.

$\rho = 5 \ \Omega\text{-cm}$

2. Since only boron has been added to the silicon, and it is an acceptor atom,

$N_A = 2 \times 10^{16}/\text{cm}^3$ and $N_D = 0$

Every acceptor atom produces one hole, so $p = N_A = 2 \times 10^{16}/cm^3$. We can solve for n since $n \cdot p = n_i^2$.

$$n = \frac{n_i^2}{p} = \frac{2 \times 10^{20}/cm^6}{2 \times 10^{16}/cm^3} = 1 \times 10^4/cm^3$$

Figure 1-7 can be used to determine the resistivity.

$$\rho = 1 \ \Omega\text{-cm}$$

3. Since arsenic atoms are donors, $N_D = 3 \times 10^{17}$ atoms/cm³. Likewise, boron atoms are acceptors, so $N_A = 5 \times 10^{17}$ atoms/cm³. Since donors and acceptors cancel each other out, $3 \times 10^{17}/cm^3$ of the acceptor atoms cancel the effect of $3 \times 10^{17}/cm^3$ donor atoms. This leaves only 2×10^{17} acceptor atoms free to give holes. Therefore, $p = N_A - N_D = 2 \times 10^{17}/cm^3$. Since

$$n \cdot p = n_i^2, n = \frac{n_i^2}{p} = \frac{2 \times 10^{20}/cm^6}{2 \times 10^{17}/cm^3} = 1 \times 10^3/cm^3$$

4. From the text,

$$R_s = 4.5 \frac{V}{I} = (4.5) \frac{5 \times 10^{-3} \text{ Volts}}{4.5 \times 10^{-3} \text{ Amps}}$$

$$R_s = 5 \frac{\text{Volts}}{\text{Amps}} = 5 \text{ ohms}$$

5. The equation for the resistance of a bar of material is

$$R = \frac{\rho L}{\text{Width} \times \text{Height}} = (2 \ \Omega\text{-cm}) \frac{100 \text{ microns}}{(5 \text{ microns}) (2 \text{ microns})}$$

$$R = \frac{2 \ \Omega\text{-cm}}{\text{micron}} (\frac{100}{10}) = 20 \ \Omega\text{-cm/microns}$$

$$R = 200,000 \text{ ohms}$$

6. Since only boron atoms have been added, and they are acceptors,

$$N_A = 5 \times 10^{16}/cm^3 \text{ and } N_D = 0$$

All of the acceptors are ionized, thus

$$p = 5 \times 10^{16}/\text{cm}^3$$

Now, we can solve for the electron concentration since $n \cdot p = n_i^2$.

$$n = \frac{n_i^2}{p} = \frac{5.9 \times 10^{26}/\text{cm}^6}{5 \times 10^{16}/\text{cm}^3} = 1.18 \times 10^{10}/\text{cm}^3$$

7. Since the intrinsic carrier concentration increases exponentially at the rate of 6% per °K, we have that

$$n_i = e^{.06\Delta T} \cdot 2.43 \times 10^{13}/\text{cm}^3$$

Now for the minority carrier concentration to be 2% of the majority concentration we may write

$$\frac{Np}{Pp} = .02 = \frac{n_i^2}{Pp^2} = \frac{(e^{.06\Delta T})^2(2.43 \times 10^{13})^2}{(5 \times 10^{16})^2}$$

Solving for $\triangle T$ yields

$$\triangle T = 94.6°$$

And

$$T = 394.6°\text{K}$$

8. Since the hole and electron concentrations are known, the intrinsic carrier concentration is thus

$$n_i = (n \cdot p)^{1/2} = (1 \times 10^{15}/\text{cm}^3 \cdot 4 \times 10^{13}/\text{cm}^3)^{1/2}$$
$$= 2 \times 10^9/\text{cm}^3$$

Finally the net impurity concentration is very nearly equal to the electron concentration or

$$(N_D - N_A) = 1 \times 10^{15}/\text{cm}^3$$

9. By electrical neutrality, 5×10^{16} donors /cm^3 must be added to the sample.

10. From the dimensions of the bar and its resistance, the resistivity can be determined from

$$R = \frac{\rho L}{H \cdot W}$$

or

$$\rho = \frac{R \cdot H \cdot W}{L} = \frac{(10\ \Omega)(0.1\ \text{cm})(0.1\ \text{cm})}{1\ \text{cm}} = 0.1\ \Omega\text{-cm}$$

Since the bar is n-type, the donor concentration is found from Figure 1.7 to be

$$N_D = 5 \times 10^{17}/\text{cm}^3$$

AV-2   SEMICONDUCTOR PHYSICS II

1.   Since the bar contains only acceptor atoms, $n = N_D = 2 \times 10^{15}$ /cm³. The hole concentration is determined using the equation:

$$p = \frac{n_i^2}{n} = \frac{2 \times 10^{20}/\text{cm}^6}{2 \times 10^{15}/\text{cm}^3} = 1 \times 10^5/\text{cm}^3$$

The formula for conductivity is:

$$\sigma = q\ (\mu_n n + \mu_p p)$$

But, since $n \gg p$, $\sigma \cong q\mu_n n$ and

$$\rho = \frac{1}{q\mu_n n} = \frac{1}{(1.6 \times 10^{-19}\ \text{coulombs})\ \mu_n\ (2 \times 10^{15}/\text{cm}^3)}$$

From Figure 2-5, $\mu_n = 1200\ \dfrac{\text{cm}^2}{\text{V-sec.}}$. Therefore,

$$\rho = \frac{1}{(1.6 \times 10^{-19})\ (1200)\ 2 \times 10^{15}\ \dfrac{1}{\text{volt}}\ \dfrac{(\text{coulomb})}{\text{sec}}\ \dfrac{\text{cm}^2}{\text{cm}^3}}$$

$$= \frac{1}{(1.6)(1.2)(2) \times 10^{-1}} = 2.6 \ \Omega\text{-cm}$$

From Figure 1-7, we have $\rho = 2.6 \ \Omega$-cm, so the comparison is quite good.

2.   In the silicon bar, $N_D = 3 \times 10^{18}/\text{cm}^3$ and $N_A = 1 \times 10^{18}/\text{cm}^3$. Therefore,

$$n = N_D - N_A = 2 \times 10^{18}/\text{cm}^3$$

$$p = n_i^2/n = \frac{2 \times 10^{20}}{2 \times 10^{18}} = 1 \times 10^2/\text{cm}^3$$

The total impurity concentration $C_T = N_A + N_D = 4 \times 10^{18}/\text{cm}^3$, and μ_n and μ_p can be determined from Figure 2-13 using this value for C_T.

$\mu_n \cong 170 \ \text{cm}^2/\text{V-sec}$

$\mu_p \cong 70 \ \text{cm}^2/\text{V-sec}$

$\sigma = q(\mu_n n + \mu_p p) \cong q\mu_n n$ since $n \gg p$

$$\rho = \frac{1}{\sigma} = \frac{1}{q\mu_n n} = \frac{1}{(1.6 \times 10^{-19})(170) \ 2 \times 10^{18}} \ \Omega\text{-cm}$$

e. This answer differs from that of Figure 1-7 because of the 2×10^{18} atoms that have cancelled each other out, but still modify the mobility of the carriers.

3.   Given the dimensions and resistivity of the device, we first note that for the p-region

$$N_D = 1 \times 10^{14}/\text{cm}^3$$

$$N_A \cong 6 \times 10^{15}/\text{cm}^3$$

The total number of impurities in the p-region is just

$$N = (N_A + N_D) \cdot \text{volume} = (6.1 \times 10^{15}/\text{cm}^3)(0.5 \ \text{cm})^3$$

$$= 7.63 \times 10^{14} \ \text{impurity atoms}$$

4.   conduction band
                              donor levels

                              acceptor levels
                              valence band

5. The resulting crystal is not electrically intrinsic since the mobilities of the holes and electrons decrease with increased doping.

6. a. $np = n_i^2$ is valid under equilibrium conditions

   b. $p + N_D = n + N_A$ is valid for any region in electrical neutrality.

7. The net impurity concentration is found from

$$N_A - N_D = 7 \times 10^{15}/cm^3 - 3 \times 10^{15}/cm^3 = 4 \times 10^{15}/cm^3$$

and the material is p-type.
Thus the hole concentration is

$$p = (N_A - N_D) = 4 \times 10^{15}/cm^3$$

and the electron concentration is determined by

$$n = \frac{n_i^2}{p} = \frac{(1.45 \times 10^{10}/cm^3)^2}{4 \times 10^{15}/cm^3} = 5.26 \times 10^4/cm^3$$

AV-3 WAFER PREPARATION I

1. It is easier to separate a gas from a liquid or solid than it is to separate just liquids or gases.

2. a. Silicon dioxide.

   b. The crucible is made of SiO_2, and some of it melts during the crystal growth process.

3. Silicon dioxide. The silicon dioxide crucible is capable of containing molten silicon, so it must not be molten.

4. a. The crystal orientation determines the preferred direction of wafer breakage.

   b. The break planes of the wafer are denoted by the flat ground on one edge of the wafer.

5. Polysilicon is silicon in which the atoms are not in an ordered crystal structure.

6. The inert atmosphere of argon gas during crystal growth prevents oxidation of the silicon.

7. A seed crystal is necessary to initiate the growth of the ingot with the correct crystal orientation.
8. The two variables which control the diameter of the silicon rod are pull-rate and temperature.

AV-4 WAFER PREPARATION II

1. **a.** x index $= \dfrac{1}{\frac{1}{2}} = 2$

y index $= \dfrac{1}{1} = 1$

z index $= \dfrac{1}{\infty} = 0$

b. We have a $<210>$ plane.

2. A value of $k = 1$ means that the concentration of dopant in the solid equals the concentration of dopant in the liquid.
3. The dopant with k closest to 1 will produce the flattest impurity profile. From the table, this impurity is boron.
4. The two most common orientations are (111) and (100)
5. Slip and dislocation.

AV-5 EPITAXIAL GROWTH I

1. No. As long as the crystal structure of the substrate is continued through the deposited layer, it is an epitaxial layer.
2. A misalignment of $35°$ will produce a maximum deposition rate.
3. From Figure 5-7, we see that above 4% HCl, a pitted surface will result.
4. **a.** From Figure 5-9, the maximum growth rate occurs when the mole fraction of $SiCl_4$ is .1.
   b. Silicon deposited using these growth conditions has poor crystal structure.
5. For epitaxial deposition to take place, nucleation sites and empty lattice sites must be available.
6. Nucleation sites are created by first slicing the wafers 3 to 7 degrees off-axis and then etching the wafers to expose the sites.
7. Low deposition rates and poor quality crystal structure are the disadvantages of vacuum deposition.
8. Hydrogen reduction of silicon tetrachloride

$$SiCl_4 + 2H_2 \rightarrow Si + 4HCl$$

and pyrolysis of silane

$$SiH_4 \rightarrow Si + 2H_2$$

9. Using silane at 1,050°C for 5 minutes yields an epitaxial layer of 100 microns from figure 5-10

AV-6 EPITAXY II

1. **a.** Induction heating or RF—the RF energy is coupled directly into a carbon susceptor, which heats the wafers that are lying on it.
   b. U.V.—ultraviolet radiation from special bulbs heats the susceptor by being directly absorbed.
2. The reacting species react less rapidly on a cold wall than on the hot susceptor, which means that there is less build-up on the wall.
3. **a.** Thickness
   b. Impurity concentration
   c. Crystal quality
4. $d = \dfrac{n\lambda}{2} = \dfrac{(8)\ .3\mu}{2} = 1.2\mu$
5. Epitaxial layer thickness can be determined by groove and stain or etch pit depth (see sec. 6-2).
6. For etch pits 1.838 μm on a side, the thickness of the epi layer is 1.5 μm.

AV-7 OXIDATION I

1. Quartz, a form of SiO_2.
2. Oxygen and water vapor.
3. The reaction occurs at the $Si–SiO_2$ interface.
4. **a.** Bubbler system
   b. Flash system
   c. Burnt hydrogen system
5. The nitrogen is used to reduce the volume of oxygen used during the dry O_2 cycle.
6. 98°C typically.
7. Explosion.
8. Greater than

   2 micron × 900 v/micron = 1800 v.

9. Injection of a small amount of chlorine bearing compounds will trap normally mobile sodium and increase the dielectric quality of the SiO$_2$ layer.

AV-8 OXIDATION II

1. **a.** Using Figure 8-1, the oxide thickness is .2μ or 2000 Å.
   b. Using Figure 8-2, the oxide thickness is .15μ or 1500 Å.
2. No. The oxidizing species must diffuse through a layer of SiO$_2$ before it can react. If the oxide growth curve is in the transport-limited region, doubling the time does not result in a doubling of the thickness of the SiO$_2$ layer.
3. **a.** Since we are beginning with a bare silicon wafer, we can use Figure 8-1 directly. We find a value of .3μ or 3,000 Å.
   b. Using 3,000 Å as a starting point on Figure 8-2, we find that we have grown the equivalent of 9 minutes (at 1200°C in 97°C H$_2$). We add another 6 minutes, bringing the total time to 15 minutes. Figure 8-2 shows a total oxide thickness of .4μ or 4000 Å.
   c. Using 4000 Å as a starting point on Figure 8-2, we find that we have grown the equivalent of 24 minutes at 1,100°C in 97°C H$_2$O. Adding another 12 minutes brings this total up to 36 minutes. Figure 8-2 shows a total oxide thickness of .5μ or 5,000 Å.
4. In anodic oxidation, silicon is the mobile species.
5. From Figure 8-2, an additional 2 microns will require six more hours.
6. For (100) silicon at 1,100°C for 24 minutes in steam, a layer 4 microns thick will be grown. Now in dry O$_2$ at 1000°C, it would take 2 hours to grow 4 microns and 3 hours to grow 5 microns. The additional micron thus takes 1 hour to grow.
7. Steam oxidation is faster than dry O$_2$ since the water molecule is considerably smaller than the O$_2$ molecule and thus will diffuse more rapidly into more locations.
8. Transport limited implies that the number of available molecules has been limited whereas reaction rate limited implies that the temperature is the limiting factor.
9. Boron tends to be depleted from the silicon during oxidation due to its greater solubility in silicon dioxide.

AV-9 IMPURITY INTRODUCTION AND REDISTRIBUTION I

1. From Figure 9-2, we see that the solid solubility is approximately 2 × 10^{19} atoms/cm^3.

2. Again using Figure 9-2, we see that this value is approximately 1.2×10^{17} atoms/cm^3.
3. Figure 9-8 reveals that gallium has a higher diffusion coefficient.
4. Using Figure 9-8, we see that this value is 2.5×10^{-12} cm^2/sec.
5. The acceleration energy (in KeV or thousand electron volts) determines the depth.
6. During predeposition the substrate temperature (and hence the solid solubility) determines the concentration of dopant at the surface of the wafer.
7. The predeposition profile is determined by a combination of the time and temperature of the predeposition.
8. An oxide thickness of 0.12 microns will effectively mask a wafer against a boron diffusion for 1 hour and 1,100°C.
9. Seven methods of introducing dopant impurities into a silicon wafer are: solid source, liquid source, gaseous source, source wafers, chemical vapor deposition of oxide, spinning on doped oxide, and ion implantation.
10. The three variables which determine the junction depth during drive-in are the predeposition impurity concentration, the time, and the temperature.
11. The two most frequently used measurements are sheet resistivity and junction depth.
12. The constantly changing concentration profile leads to only an average resistivity measurement.

AV-10 IMPURITY INTRODUCTION AND REDISTRIBUTION II

1. From Table 10-1:
   a. 1.49008×10^{-10}
   b. 2.06
2. From Figure 10-3:
   a. The junction will be present at $\sim.3\mu$.
   b. The junction will be present at $\sim.13\mu$.
3. From Figure 10-4:
   a. The junction will be present at 1.75μ.
   b. The junction will be present at $.45\mu$.
4. Graph data from problems 2 and 3.
5. Inversely since the resistivity decreases with increasing (Q).
6. As time progresses during a predeposition, the (Q) will increase and the resistivity will decrease.

AV-11 PHOTOMASKING

1. Yes. Positive resist is "light-softened" resist.
2. **a.** 5500 rpm
   b. $.8\mu$ or 8000 Å
3. More viscous.
4. **a.** Forced hot air—hot air is circulated through the chamber heating wafers and carrying away vapor
   b. I.R.—Infrared radiation heats the wafers evaporating excess solvents.
5. Iron oxide and silicon masks are transparent to yellow light while being opaque to intense ultraviolet. Chrome masks are very hard and resist scratching while emulsion masks minimize light reflection within the opaque regions.
6. Photolithography is the transfer of an image from the mask to a wafer through the use of photosensitive photoresist.
7. Photoresist performance is characterized by adhesion, etch resistance, resolution and photosensitivity.
8. Priming removes water vapor from the wafer and improves adhesion.
9. Spinning.
10. Time and temperature during baking.
11. Develop check verifies the quality and alignment of the photoresist pattern.
12. Soft bake allows poor quality of incorrectly aligned resist to be stripped and reapplied for exposure. Hard bake increases the adherence of the photoresist for the subsequent etch step.

AV-12 CHEMICAL VAPOR DEPOSITION

1. Hot-wall reactor—the reaction proceeds on the chamber wall as fast or faster than on the substrates. It is heated using thermal resistance heating.
2. **a.** Polycrystalline silicon
   b. Silicon dioxide
   c. Silicon nitride
3. Using Figure 12-6, the phosphorus concentration is 7×10^{20} atoms/cm^3.
4. The reaction chamber provides a controlled envelope around the reaction zone.

5. Reaction chamber
   Gas control section
   Time and sequence section
   Heat source for substrates
   Effluent handling system
6. Epitaxial growth is the special case of chemical vapor deposition during which the grown layer assumes the same crystal orientation as the substrate.
7. $3SiH_4 + 4NH_3 \rightarrow Si_3N_4 + 12H_2$

AV-13   METALLIZATION

1. See text
2. See text
3. E-beam
4. Planetary
5. Aluminum meets most of the requirements indicated on p. 120.
6. To prevent reaction and electromigration respectively.
7. The chamber, vacuum pumps and monitor instrumentation.
8. Four deposition methods are filament, E-Beam, flash and induction evaporation.
9. A typical vacuum deposition cycle includes:
   Wafer clean and dry
   Wafer load
   Rough vacuum
   High vacuum
   Evaporate source
   Deposit
   Stop source
   Backfill
   Unload wafers

AV-14   DEVICE PROCESSING: FROM ALLOY TO SALE

1. From Figure 14-3, the two compositions are:
   a. 22% Au, 78% Si
   b. 44% Au, 56% Si
2. From Figure 14-1, the composition is 11.3 atomic percent aluminum and 88.7 atomic percent silicon.
3. From Figures 14-1 and 14-3, we see that the aluminum–silicon eutectic temperature is higher.

4. **a.** Diamond scribing
   b. Laser scribing
   c. Sawing
5. **a.** TC bonding
   b. US bonding
6. The post alloy probe step provides the designer with an indication of process variations.
7. Backside lapping and backside metal deposition prepare the backside of the wafer for die attach and also remove builtup contamination.
8. Gold has excellent soldering and thermal properties.
9. Non-functional die are inked during wafer sort.
10. Typical steps include scratch protection, backside preparation, wafer sort, device separation, die-attach, wire bonding, packaging, final test, and mark and pack.

AV-15   DEVICES

1. Bipolar and MOS technology
2. Bipolar technology has seven masking steps while MOS technology has 5 MOS masking steps.
3. Shorting any two leads of a transistor together form a diode between the two shorted leads and the third lead.
4. A pinched-base resistor usually has a much higher resistance per unit area.
5. Reduced saturation resistance
6. The gate oxide isolates the gate material from the channel.
7. The dielectric capacitor offers higher breakdown voltages and larger capacitance values than the junction capacitor.
8. n-channel and p-channel transistors, resistors and capacitors can be fabricated using MOS technology.
9. NPN transistors, PNP transistors, diodes, resistors and capacitors can be fabricated using bipolar technology.

AV-16   CONTAMINATION CONTROL

1. Sodium
2. **a.** Deionization
   b. Reverse osmosis
3. **a.** Solvent clean
   b. The solvents remove organic contaminants such as waxes that may not react with the acids.

4. Ideally, the ceiling of a wafer fabrication area would consist entirely of laminar flow hoods. Unfortunately, their expense prohibits their use overall, except for the most sensitive processing stations.

5. Heat in H_2SO_4 to remove organics
Heat in aqua regia to remove metals
Dip in dilute HF to clean oxides
Rinse in H_2O to remove acids
Dry to prepare for next step.

6. See Figure 16-1

7. The major differences are the p^H adjustment, the filtering and the use of reverse osmosis.

8. Inert plastic piping

9. Copper tubing for oxygen and nitrogen and stainless steel for others.

10. See text.

AV-17 ADVANCED SILICON TECHNOLOGY

1. Edge-defined film-fed growth of EFG

2. **a.** Electron beams
   b. X-ray beams

3. Silicon oxidizes much more rapidly than silicon nitride.

4. I^2L devices use a common n^+ layer as an emitter, so they do not have to be isolated.

5. Lower processing cost per circuit. The size of the ingot.

6. Between 0.5 and 1.0 microns

7. Between 2.5 and 5.0 microns.

8. No mask to wafer contact and no final mask.

9. See Figures 17.4 and 17.5

10. Extremely simple structure and thus very few processing steps.

AV-18 NONSILICON TECHNOLOGY

1. An LED display requires more power because it actually emits light as opposed to just reflecting or transmitting it.

2. Quartz is a piezoelectric material which means that it vibrates when a voltage is applied.

3. Thin-film hybrid circuits use thin layers of vacuum-deposited material while thick-film hybrid circuits are formed by screening a layer of paste on a substrate and firing it at an elevated temperature.

4. Epital growth
Surface oxidation
Photolithographic transfer
High temperature diode diffusion
Ohmic contact formation
Interconnect
Test and pack

Index